CHICAGO PUBLIC LIBRARY
BUSINESS / SCIENCE / TECHNOLOGY
400 S. STATE ST.   60605

MONOGRAPHS IN MICROSCOPE SERIES

List of titles

Five volumes have now been published. Other titles will be published over the next few years. A complete listing follows:

1. A Short History of the Electron Microscope
2. Modern Electron Microscopes (SEM, TEM: design, applications, limitations)
3. Accessories for the Transmission Electron Microscope (stages, apertures, cameras, image enhancement SAD, SSE etc.)
4. Preparation of Samples and Other Techniques for the Transmission Electron Microscope (sectioning, staining, replication etc.)
5. Scanning Electron Microscopy (sample prep, use of SEM, SSD etc.)
6. Specialized Electron Microscopes (emission, reflection, high and low voltage)
7. Field-Ion Emission Microscopes (Mueller's work, one-atom probe)
8. X-Ray Microscopy (projection, microradiography, Kirkpatrick)
9. Microprobes (instruments, electron, ion, mini: design, maintenance, operation)
10. Microprobes (specimen preparation, techniques, automation)
11. A Short History of English Microscopes (mechanical design emphasized)
12. A Short History of American Microscopes (mechanical design emphasized)
13. A Short History of Light Microscopy (techniques, top lighting, polarized light, darkfield, thermal analysis, apochromats, fluorescence, interference, phase, dispersion staining)
14. The Optical Performance of the Light Microscope, Part I
15. The Optical Performance of the Light Microscope, Part II
16. Accessories for the Light Microscope (mechanical stages, micromanipulators, Lieberkuhn, micropolychromar, dispersion staining, demonstration ocular, hot stages, cold stages, drawing cameras, reticles, fiber optics imagery, DTA, stereoscopy, microprojection)
17. Special Methods in Light Microscopy (increase resolving power, increase specimen contrast, sample characterization, specimen preparation, microscopy as adjunct to other techniques)
18. Photomicrography (stereo, Cine, serial section Cine)
19. Photomacrography
20. Polarized Light Microscopy (transmission)
21. Polarized Light Microscopy (reflection)
22. Metallographic Techniques
23. Crystal Morphology
24. Microscopy in the Ultraviolet
25. Microscopy in the Infrared
26. Microspectrophotometry (absorption and emission)
27. Holographic Microscopy
28. Phase Microscopy
29. Interference Microscopy
30. Fluorescence Microscopy
31. Microtomy
32. Sections of Hard Materials (thin and polished)
33. Clinical Microscopy
34. Microphotography
35. Dispersion Staining
36. Thermal Microscopy 1-component system
37. Thermal Microscopy 2-component system
38. Micrometry
39. Stereology
40. Automatic Image Analysis
41. Mineral Identification (thin sections)
42. Mineral Identification (grains)
43. Microchemical Tests
44. Characterization of Single Small Particles
45. Study of Fibers
46. Study of Surfaces
47. Resinography
48. The Microscopy of Liquid Crystals
49. Universal Stage
50. Integration of Microscopy into the Research Laboratory
51. Dictionary for Microscopy
52. Teaching Microscopy
53. Image Processing
54. Microscopy for Art Conservators
55. Microscopy for Parents
56. Modern Optical Methods in Microscopy

SPECIAL METHODS IN
LIGHT MICROSCOPY

SPECIAL METHODS IN
LIGHT MICROSCOPY

Robert B. McLaughlin
211 Sereno Drive
Santa Fe, New Mexico 87501

Microscope Publications Ltd.
London, England
Chicago, Illinois

1977

U.S. copyright © 1977 by Microscope Publications Ltd.

ALL RIGHTS RESERVED

No part of this work may be reproduced or utilized in any form or by any means: electronic, mechanical or optical, or by any information storage or retrieval system, without the written permission of the publishers except one copy for personal use by the purchaser.

Microscope Publications Ltd.

28 Southway
Carshalton Beeches
Surrey
England

or

2820 South Michigan Avenue
Chicago, IL 60616 U.S.A.

Library of Congress Catalog Card Number: 77-86749

ISBN No. 0 904962 06 7

Printed by Newnorth Artwork Ltd. Bedford England

Dedicated to my wife Katharyn, daughter Susan, and son Scott, for their constancy of support and forbearance.

——about the author

R. B. McLAUGHLIN

——received his bachelor's degree in electrical engineering at Tri-State College in Angola, Indiana in 1949. He was employed at the time of his recent retirement as Assistant Chief, Electronics Engineering Branch, Federal Aviation Administration, Anchorage, Alaska.

From 1949 to the present he has carried on the study and practice of microscopy. His major field of interest is in the diatomaceae. He is the author of Volume 16 of this series.

He is a member of the New York Microscopical Society, the Postal Microscopy Society, the Quekett Microscopical Club and a Fellow of the Royal Microscopical Society. In 1977 he was appointed a Research Associate of the McCrone Research Institute.

PREFACE—INTRODUCTION

The title of this book may imply to some that unusual or extraordinary methods of light microscopy are treated. That is not the case. The term "special methods" is used to indicate limited, particular, specific or selected methods of light microscopy.

Topics included herein are not new or unusual at all, most having been subject matter in numerous past publications, and will, no doubt, be included in many future publications.

The subject matter of the first two chapters, resolution and contrast, has been examined, discussed and treated both theoretically and practically from the early days of light microscopy. The remainder of the book also includes material that is not particularly new, although an attempt has been made to include current information.

Microscopy is not a "cookbook" process and those who would become more than uninformed or casual users of the instrument must acquire some technical background and familiarity with both theoretical and practical aspects of it.

Volumes 14 and 15 of this series provide a sound, though elemental, basis on the theory of the light microscope. It is intended that this volume provide the more practical follow-on aspects of adjustment and use, and, in addition, furnish an overview of some of the many techniques with which the light microscope is allied.

The practice of microscopy has its biases, personal viewpoints and prejudices in adjustment and technique. One book, such as this, cannot resolve all of the differences nor serve all of the purposes inherent in the practice of light microscopy. Only wide reading and assimilation of many viewpoints can approach this ideal.

I have attempted to include information that is both useful and interesting. Wherein certain possibly controversial topics are touched upon, I have tried not to be dogmatic. When it seemed necessary to provide some basic theory for a better understanding of an adjustment, procedure or technique, I have done so. On the other hand, many of the topics covered in this book are covered at book length in other volumes of this series. Therefore, many of them are presented "bare-bones", and it is assumed that the reader who wishes to obtain a detailed treatment of a specialized subject will avail himself of information contained in references listed in this book, or in companion volumes of the series. The choice of what to include and where the emphasis has been placed has been mine. Any shortcomings, errors and/or oversights are entirely my responsibility.

Manufacturers of microscopes and associated equipment have been most generous in supplying photographs and other illustrations for my use. Many have provided technical papers, catalogs and other data and permitted me to include such information either verbatim or in abstracted or paraphrased form. Carl Zeiss Incorporated has supplied valuable information across the spectrum of light microscopy. Others that have contributed greatly to the content of this book are: Walter C. McCrone Associates, Inc., American Optical Corporation, Bausch & Lomb, Nippon Kogaku K.K. (Nikon), Unitron Instrument Company, Gaertner Scientific Corporation, R. P. Cargille Laboratories, Modulation Optics, Inc., M.E.L. Equipment Co. Ltd. (Watson, the Microscopy Division), Graticles Ltd., ATM Corporation, Universal Optics Ltd., Lemont Scientific, Olympus Corporation of America, and North American Philips. Where appropriate, credit is given for photographs, drawings or adapted drawings in the figure captions.

Finally, I wish to express my appreciation to Dr. McCrone for his assistance and continuing confidence in me, and to his staff for their unflagging support.

Robert B. McLaughlin

VOLUME 17

SPECIAL METHODS IN LIGHT MICROSCOPY

by Robert B. McLaughlin

Table of Contents

Preface/Introduction	vi
Chapter 1. Methods in Achieving Improved Resolution	1
A. Resolution - A definition	1
B. Resolving power formula - A guide	1
C. Choosing optics	3
1. Objectives	3
2. Oculars	19
3. Substage condensers	26
D. Equipment adjustment	30
1. Illumination	30
2. Substage condenser	37
3. Coverslip thickness	41
4. Immersion oil	51
5. The use of the objective back focal plane	54
6. The binocular microscope	59
E. Working conditions	60
1. Relative lighting	60
2. Cleanliness	60
F. Optical artifacts and spurious resolution	61
1. Optical artifacts	62
2. Other optical artifacts	63
G. Microscope testing	65
1. Parameter measurements	65
2. Test objects	68
H. Summary	76
I. References and commentary	76
Chapter 2. Methods in Achieving and Improving Contrast	81
A. Contrast - Fundamental considerations	81
1. Visibility	81
2. Intensity contrast	82
3. Color contrast	84

	B.	Control of the substage diaphram	86
		1. Iris diaphram	86
		2. Oblique illumination	87
	C.	Darkfield illumination	87
		1. Darkfield stops	87
		2. Darkfield condensers	88
		3. Practical matters	89
		4. Incident illumination	91
		5. Ultramicroscopy	92
	D.	Use of filters	93
		1. Selective filters	93
		2. Liquid filters	100
		3. Infrared filters	100
		4. Polarizing filters	101
	E.	Phase contrast	102
		1. The basic system	102
		2. Practical considerations	103
	F.	Anoptral phase contrast	107
	G.	Hoffman modulation contrast system	108
	H.	Interference contrast	108
		1. Jamin-Lebedeff system	110
		2. Differential interference contrast (DIC)	114
	I.	Schlieren microscopy	121
	J.	References and commentary	123
Chapter 3.		Specimen Preparation and Observation	127
	A.	Introduction	127
	B.	Killing and fixing	127
		1. Artifacts	128
	C.	Sectioning	128
		1. Cutting	128
		2. Grinding	130
		3. Corroding	130
		4. Artifacts	131
	D.	Staining	131
	E.	Surface examination	133
		1. Grinding and polishing	133
		2. Etching	134
		3. Replication	135
		4. Semi-embedding	136
		5. Casting	136

		6.	Metallization	136
		7.	Artifacts	136
	F.	Mountants	137
		1.	Introduction	137
		2.	Properties of mounting media	138
		3.	Low index media	145
		4.	Listing of mounting media	145
		5.	Immersion fluids	145
		6.	High dispersion liquids	149
	G.	Mounting methods	149
		1.	Dry mounts	150
		2.	Resinous mounts	155
		3.	Fluid mounting	159
		4.	Special considerations	167
		5.	Glass	173
	H.	Special observing conditions	174
		1.	The portable microscope	174
		2.	The inverted microscope	177
	I.	References and commentary	179

Chapter 4. Sample Characterization 183
    A. Introduction 183
    B. Characterization by light phenomena ... 183
        1. Reflection 183
        2. Refraction 183
        3. Dispersion 186
        4. Diffraction 186
        5. Interference 186
        6. Absorption 187
        7. Polarization 187
        8. Excitation 190
    C. Polarized light microscopy 190
        1. The polarizing microscope 191
        2. Characteristics determined 193
        3. Characteristic indicators 193
    D. Interference microscopy 197
        1. Characteristics determined 197
        2. Characteristic indicators 198
    E. Fluorescence microscopy 202
        1. The fluorescence microscope 202
        2. Characteristics determined 204
        3. Characteristic indicators 204
    F. Microhardness testing 208

	G.	Characterization by form and orientation	211
		1. Chemical means	212
		2. Crystal rolling	213
		3. Rotation apparatus	214
		4. Universal stage	215
	H.	Dispersion staining	216
		1. Darkfield method	217
		2. Dispersion staining objective	220
		3. Characteristics determined	221
		4. Characteristic indicators	221
		5. Special considerations	222
	I.	Ultraviolet microscopy	223
	J.	Microphotometry	224
		1. Introduction	224
		2. Instrumentation	225
		3. Characteristics determined	227
	K.	References and commentary	229
Chapter 5.		Counting and Image Analysis	232
		1. Introduction	232
		2. Sampling	233
		3. Aids to counting	233
		4. Counting procedures	241
		5. Slide preparation	247
		6. Compositional and quality counts	250
		7. Areal and linear analysis	251
		8. Point counting	253
	B.	Particle geometry	257
		1. Particle size and shape	258
	C.	Automatic accessories	260
		1. Automatic point counter	261
		2. Differential count recorder for haematology	262
	D.	Automated image analysis	264
		1. Basic system	264
		2. Data results	268
		3. Displays and presentation	270
		4. Discussion	271
	E.	References and commentary	271
Chapter 6.		Microscopy as Adjunct to Other Techniques	273
	A.	Introduction	273

B.	Industrial operations		273
	1.	Measuring microscopes	274
	2.	The cathetometer	278
	3.	Toolmaker's microscope	278
	4.	Depth measurements and surface finish	289
	5.	Comparators	293
	6.	Extensometers	293
C.	Medical practice		294
	1.	Operation microscopy — surgery	294
	2.	Ophthalmology	297
D.	X-Ray microscopy		300
	1.	Contact microradiography	301
	2.	Projection microradiography	305
	3.	Electron microprobe	305
E.	Ion microprobe		308
F.	Laser microscope		311
G.	Holographic microscopy		313
H.	Specialized research instrumentation		317
	1.	Introduction	317
	2.	Optical research	317
	3.	The cytopherometer	317
	4.	Flow systems	318
	5.	Visual stimulation apparatus (VSA)	324
I.	References and commentary		324
Subject Index			329
Author Index			336

SPECIAL METHODS IN LIGHT MICROSCOPY

CHAPTER 1
METHODS IN ACHIEVING IMPROVED RESOLUTION

A. RESOLUTION - A DEFINITION: Resolution may be defined as the minimum separation of parallel lines or adjacent points in a given subject that can be made visible as separate lines or points in the image under actual conditions.

A key word in this definition is "visible". The human eye, its characteristics and the conditions under which it functions are very important to the ability to perceive detail. In critical light microscopy the microscope, the eye and the brain function as a system.

To achieve the ultimate use of the light microscope, the microscopist must not adhere blindly to some of the more established "rules" of microscopy as immutable doctrine. On the other hand, he must accept the fact that to obtain best results from the instrument, certain fundamental considerations must be attended to.

The effect of cumulative errors with improper adjustment will prevent the best results from being obtained. There are literally dozens of factors which affect the ultimate interpretation of what is seen by the observer. While, in many cases, each has a very small effect, the cumulative result of neglecting them results in a comparatively large degree of image degradation. It has been estimated that at least a 50 percent improvement in results can be obtained by adhering to the more fundamental adjustment procedures quite often neglected completely by the practicing microscopist.

In the following paragraphs of this chapter are some of the more important considerations which, if taken into account in operation of the light microscope, will provide the best resolution of detail. Enhancement of the image by methods to improve contrast are treated in the next chapter.

B. RESOLVING POWER FORMULA - A GUIDE: The resolving power of a lens (the ability to separate detail), used under a given set of conditions, is essentially a physical limit which generally is only approached, seldom reached, and never surpassed.

There have been many equations developed for expressing the resolving power of a lens. The one that has the greatest practical value is:

$$d = \frac{1.22C\lambda}{NA} \qquad \text{Equation (1)}$$

where: d = separation distance of two self-luminous points in the object plane.
C = a factor varying between a minimum of 0.4 and 1.0 depending on such factors as the correction of the objective, and the individual capacity of observers to detect minute differences in intensity.
(lambda) = wavelength of the illuminating light.
NA = numerical aperture of the objective.
1.22 = a constant.

The equation above is developed for conditions of self-luminosity, and is valid for equivalence to self-luminosity.

If we use objectives and condensers of the highest perfection, the factor C can be assumed to be slightly more than 0.4. If this becomes the case then the equation reduces to approximately:

$$d = \frac{0.5\lambda}{NA}$$

or $\qquad d = \dfrac{\lambda}{2NA} \qquad$ Equation (2)

and the minimum separation distance to be resolved with white light of approximately 0.55 μm wavelength (λ), and an objective with a 1.25 NA is:

$$d = \frac{0.5 \times 0.55}{1.25} = 0.22 \, \mu m$$

Objectives with a slightly higher NA (1.40) and light of somewhat shorter wavelength can provide a resolution figure of somewhat less than 0.2 μm which is the generally accepted limit for the visual light microscope (Figure 1).

From Equation (1) it is obvious that an increase in resolving power (decrease in the value of d) can be obtained by a decrease in the value of C, a decrease in the wavelength of the light used and an increase in the numerical aperture (NA) of the objective. The converse, of course, is true. These, alone, comprise the significant factors on which the microscopist can base his actions to achieve the best resolution. There are other equations expressing the relationships of factors affecting the resolution obtainable, but Equation (1) has been chosen because it recognizes,

RESOLVING POWER FORMULA

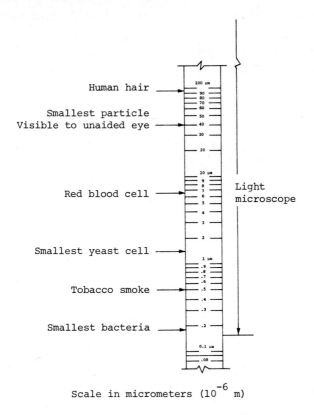

Figure 1. The range of the light microscope.

through the inclusion of the factor C, that lens aberrations and the human eye are a part of the relationship. This simple expression, then, will provide much on which to base a choice of optical elements, lighting requirements and adjustment of the system.

C. CHOOSING OPTICS: A most important factor in optimizing the resolution of the light microscope is the intelligent selection of the optical components to be used. Those components are the objective, the substage condenser and the ocular(s).

   1. Objectives: Choice of objectives for optimum resolution capability is based on numerical aperture (NA) and on aberration

corrections applied (factor C in Equation [1]) to the particular objective. Refer to Figure 2 for some typical lens arrangements.

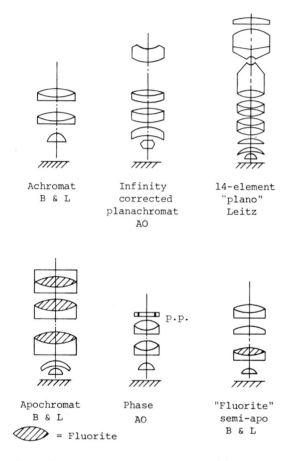

Figure 2. Lens arrangement. Some objective types.

a. <u>Achromat</u>: The achromatic objective is corrected for chromatic aberrations at two wavelengths, one in the red and one in the blue, and is fully corrected for spherical aberration at only one wavelength in the yellow-green (D line). At other wavelengths in the visible spectrum the correction for spherical abberation is good but not complete. Field curvature is present.

Correction for only two colors, and the inherent field curvature in this combination, may limit its use to visual work of a routine nature. Photomicrography (especially in color) makes more rigid demands of color corrections and flatness of field. For noncritical work, however, a careful photomicrographer, especially when working in black and white, can turn out very acceptable work with the achromat.

    b. <u>Semi-apochromat (Fluorite)</u>: Fluorite lenses are combined with glass to a limited degree to obtain a compromise between the achromat and apochromat in performance. Generally they are also corrected for two wavelengths. Field curvature is present in this design also.

    c. <u>Apochromat</u>: The apochromatic objective has several fluorite lenses in combination with glass lenses, achieving correction for chromatic aberrations at three wavelengths in the red, green and blue respectively, and for spherical aberration throughout the visible spectrum to a greater extent than with the achromat. This type of objective is usually higher in numerical aperture than those of corresponding magnification in the achromats. Typical differences are: at 10X, achromat NA 0.25, apochromat NA 0.30; at 43X achromat NA 0.66, apochromat NA 0.95. Curvature of field is also present in this type of objective. The fact that improvement in color correction and a reduction of spherical aberration is accomplished in the apochromat over the achromat does not mean that curvature of field has been reduced. Additionally, it is to be noted that apochromats are not well corrected for lateral color, and therefore require special oculars to compensate for that defect. Special designs by some manufacturers now provide objectives, termed planachromats and planapochromats, with reduction of field curvature — a matter of special interest to the photomicrographer.

    d. <u>Immersion objectives</u>: These objectives are designed for immersion into a medium of higher refractive index than that of air. There are immersion objectives for water, glycerine, oil of n_D = 1.52 and monobromobenzene.

This type of objective forms an immersion system, as opposed to a dry system wherein the objective is designed for image formation under conditions where air is adjacent to its front lens. While the dry system objectives never have a numerical aperture greater than 0.95, immersion objectives are designed with NA as high as 1.40. The result is greater light gathering power and much higher resolution. Achromats, semi-apochromats, apochromats, planachromats and planapochromats all may be obtained in the immersion design.

e. The infinity-corrected objective: It would be amiss to neglect this objective design (Figure 3) as it has become more and more popular in recent years, and now appears as standard with American Optical. This particular objective is available in all of the various corrections and applications available for the more conventional objectives.

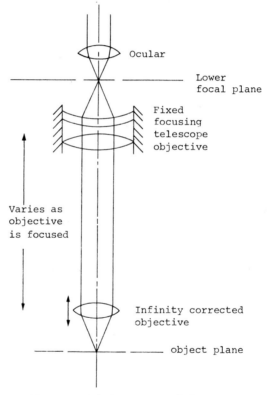

Figure 3. Infinity corrected objective.

The basic difference in the design of infinity-corrected objectives is that they emit parallel rather than convergent rays toward the ocular. Therefore, it is necessary for an intermediary fixed telescope objective to be present to focus the parallel rays on the lower focal plane of the ocular. Better peripheral corrections such as curvature of field, astigmatism and lateral color are one of the claimed reasons for the design. The improved correction ability is related to optical theory on oblique "pencils" generally connected with the foregoing aberrations. By merely shifting the aperture diaphram in design, specific deductions may be made as to the change in oblique aberrations. Thus, in infinity-corrected objective design, advantage is taken of the so-called stop-shift principle by the proper spacing of the objective elements and companion telescope objective.

While this factor is no doubt one reason for the design, it is certainly not the only one. Advantages mechanically are probably more important, in that only the nosepiece with objective need be moved, and that the intervening space between it and the telescope objective is an ideal location for accessories without the necessity for considering changes in optical path length.

The pros and cons of whether or not the infinity-corrected objective can be of a more highly corrected nature than an objective of conventional design, is not pertinent here, nor even perhaps resolvable. However, the main point is that infinity-corrected objectives cannot be used in conventional stands and vice versa.

f. <u>Comparison and choice</u>: The choice of objectives for best resolution capability rests on its numerical aperture and on the degree to which it is corrected, in design, for various aberrations. Obviously, the best objective for resolution purposes is the one which has the highest numerical aperture and the greatest degree of correction. From previous paragraphs, comparison of the corrections is obvious. To more intelligently compare numerical apertures we should at least have a brief definition and basic understanding of what that number represents.

The numerical aperture (NA) designation, usually engraved on the objective mount, expresses the resolution and light gathering properties of the objective. It is the value obtained by multiplying the sine of one-half the angular aperture by the refractive index of the medium between the front lens and the coverslip, expressed mathematically as n sine u. This important factor denoting objective resolving power is not, of course, the sole criterion of objective performance. The limit of resolution expressed or implied by this figure is exclusively <u>the distance between the smallest object points just resolved</u> (refer to Figure 4). Nothing can be said of their shape or color. The factor C in

METHODS IN ACHIEVING IMPROVED RESOLUTION

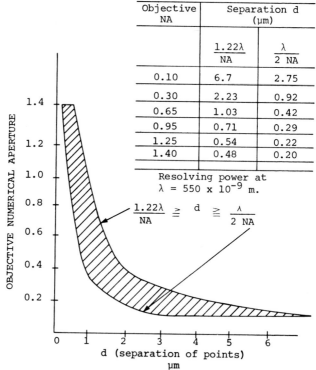

Figure 4. Resolution and numerical aperture Equation (1) from C=1.0 to C=0.4 (minimum objective correction to highly corrected objective).

Equation (1) takes the latter into account and, more appropriately then, allows the value of d to represent resolution on a "fidelity" basis, rather than merely expressing a value for separation of detail.

From the previous discussion, then, the best objectives to select wherein the numerical apertures are the same are the planapochromats since they are the most highly corrected of the group, then the apochromats, semi-apochromats and, finally, the achromats. As might be expected, as objectives become more highly corrected their cost greatly increases. The apochromats are more expensive than semi-apochromats and so on (Figure 2).

Some other inherent qualities of objectives bound exclusively

CHOOSING OPTICS

to numerical aperture and aberration correction factors are included in Table I, Table II and Figure 5. A survey of ten leading manufacturers of microscopes was made, including the United States, Europe and Asia. The numerical aperture listing in Table I represents NA's available from those manufacturers collectively. The corresponding angular aperture of such NA values are in columns 2, 3 and 4 for air, water and homogeneous immersion (n = 1.52), respectively. The heavy accented boxes in those columns indicate the actual range of NA's over which objectives are available. The remaining columns of the table indicate the theoretical resolving power at the various numerical aperture values for objectives of poor correction and construction to those of optimum correction and construction (values of C from 1.0 to slightly more than 0.4). The remaining columns of Table I indicate how the depth of field of an objective varies with numerical aperture. It might be well to very briefly discuss this last parameter.

Depth of field can be defined as the distance along the optical axis throughout which the object can be located and yet be imaged with satisfactory clarity. The definition used in calculating the figures in Table I is more exacting although perhaps not as inclusive of the human factor. That definition is: "The distance that a given subject plane may be shifted along the optic axis to let the disc of confusion equal the resolving power." Assuming two parallel lines of negligible thickness at a lateral distance equal to the limit of the resolving power of the system, the depth of field will be the displacement of the object plane relative to the objective which just causes the two lines to appear fused into one. Expressed mathematically this distance is:

$$T_f = \frac{\lambda \sqrt{n_D^2 - NA^2}}{2 NA^2}$$

and is graphically illustrated in Figure 5. It will be noted that under this definition the object plane can be displaced both above and below the plane of focus, therefore providing that visually the depth of field will be $2T_f$.

Table II has been provided to indicate the visual depth of field for some typical objective/ocular combinations. The accommodation of a good human eye is 250 mm at unity magnification. At higher magnification it decreases as:

METHODS IN ACHIEVING IMPROVED RESOLUTION

TABLE I

Numerical aperture relationships

Numerical aperture (NA)	Angular aperture (AA)				Resolution (d) (μm) $d = \dfrac{1.22 \, C \, \lambda}{NA}$				Depth of field (μm)	
$NA = n \sin u$	$AA = 2 \sin^{-1}(\dfrac{NA}{n})$ Degrees				$\lambda = 0.540 \, \mu m$ Wratten 58 + 15		$\lambda = 0.478 \, \mu m$ Wratten 45A		$T_f = \dfrac{\lambda \sqrt{n_D^2 - NA^2}}{2 \, NA^2}$	
NA	Dry $n=1$	Water 1 mm $n=1.33$	Homo 1 mm $n=1.52$		$\dfrac{1.22\lambda}{NA}$	$\dfrac{\lambda}{2\,NA}$	$\dfrac{1.22\lambda}{NA}$	$\dfrac{\lambda}{2\,NA}$	T_f	$2\,T_f$
1.40			134.17		0.470	0.193	0.416	0.171	0.082	0.164
1.32		165.93	120.55		0.500	0.205	0.443	0.181	0.116	0.232
1.30		155.63	117.56		0.507	0.208	0.449	0.184	0.126	0.252
1.25		140.05	110.65		0.523	0.216	0.463	0.191	0.149	0.298
1.00		97.52	82.28		0.659	0.270	0.583	0.239	0.309	0.618
0.95	143.61	91.17	77.36		0.697	0.284	0.617	0.252	0.356	0.712
0.90	128.32	85.17	72.60		0.728	0.300	0.644	0.266	0.408	0.816
0.85	116.42	79.45	68.00		0.779	0.318	0.690	0.281	0.473	0.946
0.80	106.27	73.97	63.52		0.820	0.338	0.726	0.299	0.572	1.154
0.75	97.18	68.65	59.13		0.881	0.360	0.780	0.319	0.638	1.276
0.72	92.10	65.53	56.53		0.912	0.375	0.807	0.332	0.695	1.390
0.70	88.85	63.52	54.83		0.943	0.386	0.835	0.341	0.743	1.486
0.66	82.60	59.50	51.47		0.994	0.409	0.880	0.362	0.839	1.678
0.65	81.3	58.51	50.63		1.015	0.415	0.898	0.368	0.884	1.768
0.60	73.73	53.63	46.50		1.097	0.450	0.971	0.398	1.047	2.094
0.57	69.50	50.75	44.05		1.158	0.474	1.025	0.419	1.191	2.382
0.55	66.73	48.85	42.43		1.199	0.491	1.061	0.435	1.276	2.552
0.50	60.00	44.17	38.40		1.322	0.540	1.170	0.478	1.550	3.100
0.45	53.49	39.55	34.44		1.466	0.600	1.298	0.531	1.961	3.922
0.40	47.16	35.01	30.52		1.650	0.675	1.461	0.598	2.474	4.948
0.35	40.97	30.52	26.63		1.886	0.771	1.669	0.683	3.330	6.660
0.33	38.54	28.73	25.08		1.999	0.818	1.769	0.724	3.640	7.280
0.32	37.33	27.84	24.31		2.060	0.844	1.823	0.747	4.015	8.030
0.30	34.92	26.07	22.77		2.194	0.900	1.942	0.797	4.470	8.940
0.25	28.96	21.67	18.93		2.634	1.080	2.332	0.956	6.750	13.500
0.22	25.42	19.04	16.64		2.993	1.227	2.649	1.086	8.116	16.232
0.21	24.24	18.17	15.88		3.137	1.286	2.777	1.138	9.235	18.470
0.20	23.07	17.30	15.12		3.301	1.350	2.922	1.195	10.172	20.344
0.17	19.58	14.69	12.84		3.875	1.588	3.430	1.406	13.590	27.180
0.16	18.41	13.82	12.08		4.151	1.688	3.674	1.494	15.691	31.382
0.15	17.25	12.95	11.33		4.397	1.800	3.892	1.593	20.430	40.860
0.12	13.78	10.35	9.06		5.494	2.250	4.863	1.992	29.218	58.436
0.10	11.48	8.62	7.54		6.591	2.700	5.834	2.390	40.950	81.900
0.09	10.33	7.76	6.79		7.319	3.000	6.479	2.656	50.573	101.146
0.08	9.18	6.90	6.03		8.241	3.375	7.295	2.988	64.028	128.056
0.075	8.60	6.47	5.66		8.784	3.600	7.775	3.187	73.189	146.378
0.07	8.03	6.03	5.28		9.420	3.857	8.338	3.414	83.656	167.312
0.0625	7.17	5.39	4.71		10.547	4.320	9.336	3.824	105.127	210.254
0.06	6.88	5.17	4.52		10.988	4.500	9.726	3.983	113.902	227.804
0.05	5.73	4.31	3.77		13.182	5.400	11.669	4.780	164.052	328.104
0.04	4.58	3.45	3.02		16.482	6.750	14.590	5.975	256.382	512.764
0.03	3.44	2.58	2.26		21.966	9.000	19.444	7.967	455.880	911.760
0.02	2.29	1.72	1.51		32.954	13.500	29.170	11.950	1025.798	2051.596

TABLE II

Visual depth of field with the light microscope

M_T	E_a (μm)	Obj.	Ocular	V_f (μm)
2000	0.0625	100X 1.25 NA	20X	0.3605
1500	0.1111	100X 1.25 NA	15X	0.4091
1000	0.250	100X 1.25 NA	10X	0.548
500	1.000	50X 0.70 NA	10X	2.486
250	4.000	25X 0.40 NA	10X	8.948
125	16,000	10X 0.25 NA	12.5X	29.50
75	44.44	10X 0.25 NA	7.5X	57.94
50	100.00	5X 0.10 NA	10X	181.90
25	400.00	2.5X 0.07 NA	10X	567.31
10	2500.0			
5	10×10^3			
1	250×10^3			

M_T = Magnification (Obj.) x Magnification (Ocular)

E_a = Eye Accommodation = $\dfrac{250 \times 10^3}{M_T^2}$ μm

$V_f = 2 T_f + \dfrac{250 \times 10^3}{M_T^2}$ (μm)

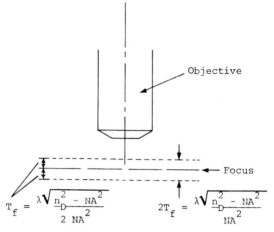

$$n_D = \text{refractive index of medium}$$

Figure 5. Depth of field.

$$\frac{250 \times 10^3}{M_T^2} \mu m$$

where M_T is initial magnification (obj.) times the magnification (ocular). Since the eye can accommodate somewhat even at high magnification, then it follows that the visual depth of field with a microscope must be that which the instrument provides plus the eye accommodation. The ability of the instrument to provide depth of field depends entirely on the wavelength of the light used, the numerical aperture of the objective and the refractive index of the medium in which the object is mounted. The eye accommodation on the other hand is related only to the total magnification factor involved.

The depth of field figures provided here are, in fact, only approximations. Abbe, Berek and others have examined this factor extensively and all differ to a considerable degree in the values for the visual depth of field through a microscope. There are so many variables involved, both physical and physiological, that exact figures are not possible. The information provided here is mainly to provide information as to the relationships involved, rather than exact figures, and to provide some degree of magnitude to the reader.

CHOOSING OPTICS

Depth of focus, a similar consideration, is roughly defined as the distance along the optical axis throughout which the image formed by a lens is focused clearly. This concept is of greater importance photomicrographically and will not be discussed further here.

Additional factors in the choice of objectives for optimum performance have to do with properties and characteristics which are in addition to the numerical aperture and aberration corrections applied. These additional considerations are included in the factor C (Equation 1). For C to attain a value such that the equation reduces to the form of Equation (2) design parameters are not all that are required. It is true that factor C includes aberration corrections also, but if a lens is poorly constructed mechanically, has internal reflections or is misaligned optically, e.g., not set precisely at right angles to the optical axis, then the definition capability of the objective will be inferior, although designwise the numerical aperture and residual corrections may be of the highest. This condition can either obtain through faulty manufacture and poor execution of the design practically, or through damage to the objective in handling or other treatment. These factors are the ones we seek in making comparisons and choices between objectives with identical numerical apertures and corrections. It is a quality consideration that used to be termed "defining power" in the early days of microscopy and still deserves attention. Comparisons for choices of objectives on this basis can only be made by test. Some simple tests for this and other quality determinations are described later in this chapter.

Other characteristics that will affect the use of objective lenses are working distance, nonpolarizing characteristics and so on. Figure 6 illustrates actual objective barrel markings by various suppliers. Some of them provide more complete information than others. Factors important for comparison purposes or for use-selection are (not necessarily in order of importance):

1. mechanical tubelength,
2. numerical aperture,
3. initial magnification,
4. aberration correction, and
5. coverslip thickness.

There are only two mechanical tubelengths (Figure 7) in general use at present, 160 mm and 170 mm. Almost all manufacturers throughout the world use 160 mm mechanical tubelength except E. Leitz, who uses 170 mm. Coverslip thickness for

METHODS IN ACHIEVING IMPROVED RESOLUTION

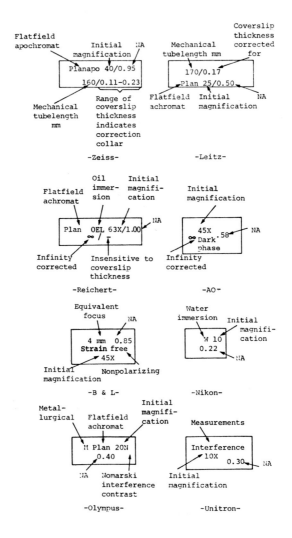

Figure 6. Objective barrel notations.

CHOOSING OPTICS 15

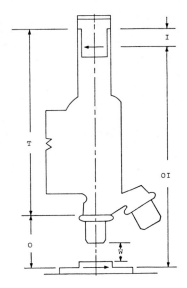

T: Mechanical tubelength
I: Intermediate image distance
O: Object distance of objective
OI: Object to image distance
W: Working distance

Figure 7. Microscope. Optical-mechanical relations. (adapted from Zeiss "Optical Systems for the Microscope".)

which spherical aberration corrections are applied in design are also of two values, 0.17 mm and 0.18 mm. The U.S. and English manufacturers almost exclusively use 0.18 mm. Most other foreign makers, including Asia, use 0.17 mm.

The most complete information actually engraved on the objective barrel is furnished by Zeiss. Others grade down from there in the detail they provide. For instance, almost all U.S. manufacturers omit the coverslip thickness designation, and almost all manufacturers do not designate achromats as such, it being "understood". Initial magnifications on the barrel are in almost any case only approximations. For a Zeiss objective marked 40X, for instance, the actual initial magnification may be

40.8X, 39.6X/or etc.

Some manufacturers provide the equivalent focal length of the objective either in addition to the initial magnification or by itself. It also is an approximate figure; a 4 mm equivalent focal length may turn out to be 4.5 mm. The equivalent focal length divided into the mechanical tubelength provides a rough approximation of the initial magnification. For instance, an objective with a barrel marked 4 mm with a 160 mm mechanical tubelength will have a normal initial magnification of 40X. However, actual values may be 4.5 mm and an optical tubelength of 183.6 mm (Figure 7). Initial magnification is actually determined by dividing the equivalent focal length into the optical tubelength. This latter dimension is variable from objective to objective and from manufacturer to manufacturer, and is dependent upon the lens assembly dimensions and spacing within the objective barrel and the location of the lower focal plane of the ocular. Some relationships of these factors are illustrated in Figure 8 for an actual Zeiss objective.

Various types of abbreviations and acronyms are used to designate flatfield optics, apochromats, semi-apochromats etc. The word "plan" or abbreviation "pl" usually refers to a flatfield objective. However, it is difficult to generalize in the matter of ocular and objective labels as all manufacturers seem to prefer their own labeling schemes. The user, if he is to mix objectives on a stand, must be very familiar with the nomenclature of those he is choosing, or he may very well degrade the results he obtains. The very obvious differences, of course, are the tubelength (160 or 170 mm), and the coverslip thickness (0.17 mm or 0.18 mm). Others not so obvious might include the fact that the lower focal plane of the ocular for a 160 mm mechanical tubelength might be different (more of this later). The objective might be corrected for use without a coverslip, or the objective may be only usable with a 200 mm mechanical tubelength, which might be the case with certain "unified" stands etc. There is no substitute for knowing the objective characteristics.

In addition to the basic objective characteristics which must be known to be used properly and to furnish the best results, there are others that deserve mention. Flatness of field is one of them. As the visual use of a microscope is inherently connected with the physiological use of the eye and the comfort and attitude of the observer, this feature deserves serious consideration, especially wherein long-term observations are concerned, and where such flatfield attributes are not gained in design

CHOOSING OPTICS

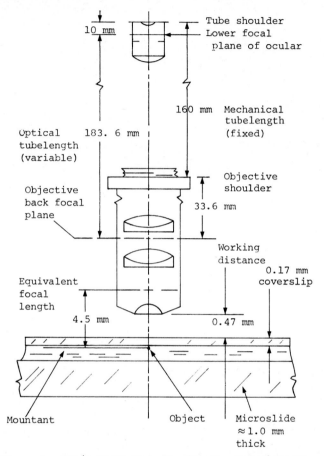

Figure 8. 40X/0.65 NA high-dry objective (initial magnification = 40.8X). No scale.

through sacrifice of other necessary corrections for maximum definition and resolution.

Special designs of objectives are numerous. Among them are phase contrast, strain-free, incident light, ultraviolet, metallurgical, long working distance and immersion objectives for water, glycerine etc. Special purpose objectives should be used for purposes for which they were specifically designed. They can be, in some instances, used for general applications or other

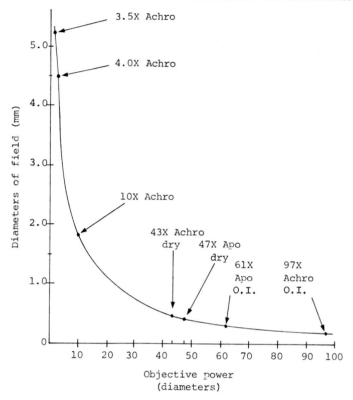

Figure 9. Viewing field diameter.

special tasks, but the user must be aware that unless he knows their characteristics to be suitable, they may derogate his image to uselessness.

Additional factors in the choice of objectives for optimum performance are mechanical. If the objectives used, or chosen from, are manufactured or provided for the stand in use, there is generally no choice problem. However, if the objectives are not for the stand in use there are some factors that must be considered for best performance.

The discrepancies among manufacturers of various makes, of tubelengths and ocular focal plane positions, for instance, can be quite great. Even with the common mechanical tubelength of 160 mm, the location of the lower focal plane of the ocular below the upper rim of the tube ("I" in Figure 7) may be several millimeters. Typical distances to this plane range from 10 to 15 mm for major manufacturers and sometimes greater for others. Not only can this type of difference affect the parfocality of the system, it may reduce the effectiveness of the objective corrections since any objective is corrected for one set, and one set only, of conjugate foci. In addition, although certain standards are adhered to as to the type of fastening (RMS thread etc.) the length of the objective barrel and the placement of the lens elements within it may vary. From these considerations, it is unwise to utilize "mixed" objectives of various manufacturers and expect their performance to be optimal unless the extant mechanical condition is identical to those for which they were designed.

A choice among types of immersion objectives may be of some importance also in realizing the most in resolution. For objects mounted in water or in aqueous media of approximately the same refractive index as water (1.33) such as blood serum, normal salt solution, seawater, plant sap, dilute sugar solution, diluted acetic acid or watery jellies, water immersion systems should be considered. Under these circumstances, except with objects known to be in optical contact with the coverslip (or within a few micrometers of it), a good water immersion objective, properly adjusted, gives markedly better, less glassy and more realistic images than those given by an equally good oil-immersion objective; the difference increasing from zero with increase in the depth below the coverslip of the watery liquid focused through.

It is not possible to indicate objective choices for all of the many requirements that must be met in microscopical work. However, if the microscopist will take sufficient time to evaluate the situation at hand in terms of resolving power required, type of specimen, mounting media and illumination conditions, his choice of a proper and adequate objective will more than likely be an intelligent and satisfactory one.

2. <u>Oculars</u>: The microscope ocular cannot improve the resolution of detail that is presented to it by the objective. However, it can degrade that image in a number of ways. For best presentation to the eye or other recording surface (photographic film etc.) the choice of the ocular may very well decide the resolution at that point. Various types of oculars are currently available. The choice of what to use for best resolution rests on some knowledge of their characteristics.

a. <u>Huygenian</u>: The Huygenian design (Figure 10) utilizes two lenses made of crown glass with a focal length ratio (of the field lens to the eyelens) of 1.5 to 3.0. A field stop is often located at the primary focal point of the eyelens which lies between it and the field lens, and it is this field stop location where cross hairs or reticles are mounted. Although this ocular design as a whole is corrected for lateral chromatic aberration, the in-

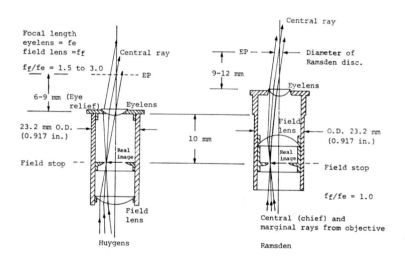

Figure 10. Comparison of oculars.

CHOOSING OPTICS

dividual lenses are not so that cross hairs or reticle scales as seen through the eyelens alone may show considerable distortion and color. Usually reticles used with Huygenian oculars should be confined to the center of the field to minimize distortion effects. This design has some spherical aberration, astigmatism and a rather large amount of longitudinal color and pincushion distortion. Also, in general, the eye relief (the distance between the top of the eyelens and the exit pupil) is comparatively short, ranging from about 6 to 9 mm.

b. <u>Ramsden</u>: In the Ramsden ocular design (Figure 10), the two lenses are usually made of the same type of glass, but they have equal focal lengths. For best lateral color their correction separation should be equal to their focal length. Since the first focal plane of the system coincides with the field lens, the reticles must be located there. This may be desirable in some cases, but any dust particles on the lens surface will also be seen in sharp focus and that is undesirable. To overcome this latter difficulty the lenses are usually moved a slight distance closer together than the ideal, moving the focal plane forward, thus sacrificing some lateral achromatism. This design has more lateral color than a Huygens, but the longitudinal color is only about half as great. It has about one fifth the spherical aberration, half the distortion and no coma. It also has the important advantage of about 50 percent greater eye relief than the Huygens.

c. <u>Compensating</u>: Compensating oculars are usually based on one or the other of the above designs, dependent upon the magnifications. General construction of the low-power compensating oculars follows that of the Huygenian design, excepting that the eyelens is a doublet. This provides correction for lateral color by causing red and blue rays to emerge parallel so that they unite in a single color-free image on the retina of the eye.

The higher power compensating oculars are of a more complex form. The Kellner ocular is essentially an achromatized Ramsden (Figure 11), and variations of this design are sometimes used in the higher power compensating oculars. There are a number of other designs for these oculars, but the significant characteristic is that they compensate for the lateral color aberrations of the higher powered objectives.

d. <u>Comparison and choice</u>: With low-power objectives

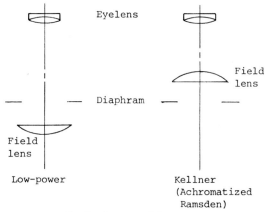
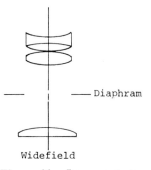

Figure 11. Lens arrangement, some ocular types.

the Huygenian ocular performs quite well and because of its inexpensive construction is the most used of all designs. However, when higher powered objectives are in use, even achromats, the compensating oculars can produce an effectively better image, resolutionwise, because they can correct some residual color aberrations present in those objectives. For the apochromat, it is mandatory that a compensating ocular be used as in its design it is left to a compensating ocular to correct for lateral color aberration. As the degree of residual lateral color is a matter of individual manufacturing design and, for other reasons, an apochromat should be paired, for best performance, with its matching compensating ocular.

Another important characteristic of the ocular, which affects

CHOOSING OPTICS

the over-all resolution, is the position and diameter of the "eyepoint" or Ramsden disc.

The Ramsden disc is the "exit pupil" of the microscope and is defined as the image of the aperture stop formed by the optics on the image side of the stop. The diameter of the exit pupil is intimately connected with the attainment of the full resolving power of the microscope. The height of the eyepoint (Ramsden disc) above the top of the eyelens depends upon the focal length of the ocular and its diameter depends on the effective numerical aperture of the objective and the total magnification of the microscope.

Considerable theoretical background on the physiological and physical aspects of the human eye is necessary for complete understanding of these relationships and is beyond the scope of this book. However, a few factors of importance to achieve best performance from the microscope and to guide the choice of oculars are presented without that background.

The exit pupil (Ramsden disc) is best restricted to a maximum of about 1.5 mm in diameter (for the normal eye) to allow eye movement, reduce aberrations of the eye and to permit full utilization of the microscope optics. (The normal eye pupil diameter under normal lighting conditions is about 3 to 4 mm.)

An optimum diameter of the Ramsden disc is about 1/3 to 1/4 the diameter of the observer's pupil. The diameter of the Ramsden disc is expressed as:

$$\frac{500 \times NA}{\text{Total magnification}}$$

or: $$\frac{500 \times NA}{(\text{ocular mag.}) \times (\text{obj. mag.})} \qquad \text{Equation (3)}$$

If in (3) above, the factor:

$$\frac{500 \times NA}{(\text{obj. mag.})}$$

is determined for each objective to be used, then the diameter of Ramsden disc can be calculated by dividing it by the ocular magnification, or:

$$\text{Ramsden disc diameter in mm} = \frac{\frac{500 \times NA}{\text{obj. mag.}}}{\text{ocular mag.}}$$

A Ramsden disc diameter of about 1 mm allows freedom of eye movement and assures that the entire disc is included within the eye iris.

If we desire this disc diameter, then for an objective of 40X and 0.65 NA the ocular magnification would be

$$\frac{500 \times NA}{obj.\ mag.} = \frac{500 \times 0.65}{40}$$

or 8.13X (an 8X or 10X ocular would be used)

Under certain high intensity lighting conditions and/or aging of the eye, for instance, the pupil diameter may be considerably less than 3 mm diameter, even perhaps as low as 1.5 to 1 mm. To improve conditions for the eye, the Ramsden disc should be made as small as comfort in viewing will allow. As the above simple equation shows, this requires the use of higher powered oculars.

High magnifications and high illumination will improve color evaluation also. In any case, where the eye is below normal in color differentiation, brightness sensitivity or form recognition, higher powers are called for and, in general, should be in use as normal practice by all microscopists.

The height of the Ramsden disc above the eyelens is greater with oculars of the Ramsden design, for instance, than with the Huygenian. As the magnification of the ocular increases, the eyepoint height decreases, uncomfortably so, in the Huygenian design.

All ordinary oculars are spherically undercorrected for the axial pencil of rays, and therefore their oblique (rays) aberrations vary when the aperture diaphram is shifted along the optical axis, and there can be only one position which gives the best possible correction. In the microscope, the objective acts as an aperture diaphram for the ocular, therefore there will be a particular "tubelength" at which any ocular gives its best results. This confirms the good sense in not mixing even oculars with equipment other than that for which they were designed if optimum results are to be obtained. Of course, if it can be determined that optical and mechanical parameters are identical, it will make no difference. Unless that assurance can be obtained, best performance is in doubt.

Whether to wear eyeglasses (if they are worn normally) in using the microscope can be determined quite easily by holding them to view a symmetrical object and then rotating them to determine if the object changes its symmetry. If it does, corrections have been applied to them that are not taken care of by the microscope optics and they should be worn when using the instru-

CHOOSING OPTICS

ment. Simple axial focusing problems corrected by eyeglasses can be taken care of by focusing action of the microscope and eyeglasses need not be worn under that condition. If eyeglasses are necessary, the eyepoint of the ocular must be higher to accommodate the placement of the eyeglass lens. Specially designed oculars are available, with high eyepoints (as much as 20 mm above the eyelens) for that purpose.

    e. Total magnification: The "rule of thumb" whereby the maximum total magnification of a microscope should not exceed 1000 times the NA requires some comment. There are those who even propose that 500 to 750 times the NA is sufficient as a criterion for selecting ocular-objective combinations.

The *ideal* human eye, in *rare cases,* under *optimum* conditions of illumination and contrast, can distinguish objects subtending less than 40 seconds of arc (visual angle) or about 40 μm at a normal accommodation distance of 250 mm. However, the dependence of the eye on illumination and contrast for maximum resolution is very great. Under optimum conditions of contrast, the good or average eye is only about one-fourth as good at resolving as the "ideal eye". Assuming the good or average eye to resolve objects separated by about 150 μm under optimum illumination and contrast conditions, the magnification required by the microscope optics is of interest.

Assuming a wavelength for the illuminating light of 500 nm and an NA of the objective to be 1.25, the nearly maximum resolution is 0.20 μm. The total instrument magnification required for the eye to distinguish that separation at its 250 mm accommodation distance (as it is at the ocular) is:

$$\frac{150}{0.20} \text{ or } 750X$$

It is evident that the magnification is less than 1000 NA (1000 x 1.25), but certainly is greater than 500 NA. Also, thus far, we have been considering a good or average eye under *optimum* image contrast conditions. It is not at all exaggerated to say that the usual contrast conditions with a light microscope are less than optimal. The eye, under such instrument conditions (or the unassisted eye for that matter), at the lower limit of image contrast, can only resolve separations about 4 times as great as under optimum condtions. That means that for the example we have been discussing the total magnification would have to be 4(750) diameters or 3000 diameters. This is indeed much greater than 1000 NA (1250X), being more than twice that dictated by the "rule". The "rule" was based, by Abbe, on the resolution of certain periodic structures under optimum contrast conditions.

Some slight self-evaluation by the average microscopist as to how well he can distinguish fine periodic structure under optimum conditions and adequate illumination at his desk will be very revealing of his resolution capability. Examination of half-tone screen prints ranging from 1/100 inch to 1/300 inch (254 μm to 88 μm) will illustrate the point.

Further, the details of microscopic specimens examined are rarely periodic in nature and usually not black and white, but often in color, and contrast is usually rather low.

From this very brief discussion it is apparent that the 1000 NA rule is not usually interpreted properly as was intended by Abbe. In fact, with the average or below average eye, it should be the starting point for magnification determination, rather than the maximum limit. The extensive discussion by C. Van Duijn in The Microscope (referenced at the end of this chapter) is a convincing argument for acceptance of this concept. It cannot be recommended too highly and all microscopists who have not read and studied it who are interested in obtaining the best from their instruments should do so.

Because of the vagaries of the human eye, illumination and contrast, and types of specimen material examined, it is not possible to make positive statements regarding the degree of benefit of higher than 1000 NA total magnification, but, in the same sense, the limitation imposed by this "rule" has restricted, in many instances, the best use of the instrument. Personal experience with higher than 1000 NA magnifications will be the criterion for judgment, and will prove to the user to be more beneficial than detrimental.

Under conditions of less than optimum contrast, start with 1000 NA magnification and increase from that point to magnification selection for best visibility and resolution under the particular condition at hand. Other advantages of high-power magnification will be covered later in this chapter.

3. Substage condensers: Aside from special types of substage condensers there are two or three designs available for general transmitted light use. Their characteristics are briefly outlined below.

    a. The Abbe condenser: The Abbe condenser is available in both a fixed and modified variable focus form. The latter allows better illumination of large fields in low-power work.

CHOOSING OPTICS

However, both forms are similar in optical performance. The lens system is composed of a plano-convex hemispherical upper element and a biconvex lower element (Figure 12).

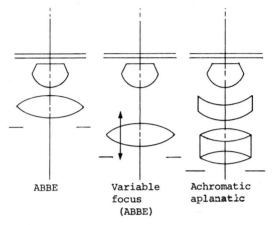

ABBE   Variable   Achromatic
              focus   aplanatic
              (ABBE)

Figure 12. Condenser types.

Considerable aberrations are inherent in this system, not the least of which is spherical aberration. Since the Abbe condenser is not corrected for spherical aberration, it generally does not live up to its rated NA. In order for it to do so would require a very large light source which, of course, produces considerable haze and obscuration of detail. To eliminate the haze, the light source must be reduced in size, and that then produces the undesirable effect of a strongly marked caustic (an area of nonconvergent light rays) due to uncorrected spherical aberration. Thus, because of this one uncorrected defect, the Abbe condenser is not suitable for critical work. In addition to spherical aberration it also suffers from chromatic aberration. However, in spite of these defects, it performs adequately for many types of routine microscopical work and has the advantage of being reasonably priced because of its simple construction. For some work, the chromatic aberration of this condenser is offset by the use of color filters. However, when color photomicrography is contemplated, the chromatic aberration of the Abbe becomes a matter for serious consideration, and a limiting factor in its use.

b. The "corrected" condenser: The shortcomings of the simple two-lens Abbe have been overcome in a number of other designs. The best of these designs are the achromatic-aplanatic condensers. The spherical aberrations are corrected by a more complex arrangement of lens elements. A truly aplanatic-achromatic condenser is corrected spherically for rays away from the axis, as well as axial rays, and correction for chromatic aberration is such that it can be used with the higher performance objectives. Construction is much more complex, and commonly consists of five or six elements of glass in cemented or air-spaced groups (Figure 12).

      c. Comparison and choice: For high power critical microscopy, the corrected condenser should be used. Not only does it provide for better focusing of diaphramed light, thus reducing glare, but it allows the more highly corrected objectives, such as apochromats, to work at their best resolving capability.

While the Abbe is useful for much routine work with the light microscope, improvement in resolution capability of the system can be obtained by using a corrected condenser. If the work is of a critical nature, or highly corrected objectives are in use, it is essential that a corrected condenser be used so as to not derogate their performance.

Aside from the specific corrections which the condenser may have, it is of utmost importance that it is capable of supplying light to the object at such an angle that the objective can attain its maximum resolving power. The objective can only do this under full-cone illumination or extreme unidirectional oblique illumination. For a 1.40 NA objective to perform at its maximum capability requires a cone of light whose solid angle is 134.16° in an immersion medium of 1.52 refractive index (Table I). Further, the condenser itself, unless immersed in the same medium (even though its "NA" is 1.40) cannot fulfill that requirement. Condensers are dry or immersion types, as are objectives, and their rated "NA" is only attained under the conditions they were designed for. The example above requires a 1.40 NA condenser oiled to the bottom of the microslide, as well as the objective being oiled to the coverslip for an effective maximum NA to be available. If the condenser is not oiled to the bottom of the microslide, then its numerical aperture is limited to 1.40 /1.52 or 0.92, and the effective aperture of the system is (1.40 + 0.92) divided by 2 or 1.16 NA, less than the NA of the objective and would limit the resolving power to 1.16 capability. A dry condenser of 0.9 NA used with such an objective would provide a system NA of (1.40 + 0.9) divided by 2 or 1.15. Generally condensers of NA higher than 0.9 should be oiled when used with an oil immersion objective.

In the dry 0.9 NA condenser example above, it will be noted that it provides the same system NA and has better correction since the correction of the 1.40 NA condenser suffers because it is not used immersed as its design calls for. In addition, the reduction in NA by this means for visual work is appropriate, wherein the rays are not traversing the peripheral portions of the objective, where aberrations are least able to be corrected. The 0.9 NA condenser, in this case, is furnishing a decreased illuminating cone of approximately the size to provide improved quality of the image; much the same as if the 1.40 NA condenser were used oiled, objective oiled, and then the substage diaphram stopped down to offset peripheral aberrations of the objective.

The free working distance of a substage condenser is of considerable importance, especially if extra thick microslides or special flasks or containers are to be used for specimens on the microscope stage. Condensers can be obtained with working distances ranging from as high as 37 mm to about 1.8 mm to 2.5 mm for nominal thickness slides. Sometimes called the object distance, it is the position of an image of an infinitely distant lamp field stop above the stage surface in glass for a medium aperture and a condenser racked up to its upper stop. It is a measure of the thickness of the objects which can be properly illuminated with the condenser by the Köhler method.

The equivalent focal lengths of condensers vary between about 8.4 mm and 13.0 mm for normal object distances (1.6 - 2.2 mm), and somewhat shorter for the highly corrected achromatic/aplanatic types. Many condensers feature a removable top lens or a "swing-out" top lens so that full illumination may be furnished to the lower power objectives. The removal or "swinging out" of the top element converts the condenser lens system from a short focus to a long focus one, and a consequently lower maximum aperture.

It is to be noted that quite often "achromatic" condensers have also been corrected for spherical aberration and then can be considered the equal of the achromatic/aplanatic types. The term "aplanatic" generally indicates corrections for both spherical aberration and coma. The term aspheric indicates a lens not ground on a spherical basis, such that the rays of light from the marginal zones focus at the same point as those through the center. Condensers so designated are not necessarily achromatically corrected, however.

Condensers with a decenterable aperture iris diaphram are still available from some manufacturers in the achromatic/aplanatic types, although this feature is not as much in demand as in years gone by. The decentering of the iris diaphram allows considerable contrast control and is sometimes very useful, especially when other contrast methods are not available.

The "Pancratic" condenser of Zeiss deserves mention. It is a somewhat "automated" form in that as the illuminating aperture is adjusted to suit the objective in use, a corresponding reduction in the size of the illumination is accomplished, maintaining a fixed product of the field diameter and aperture. This is accomplished by a single control which changes the focal length of the optical system comprising the condenser. It is an achromatic/aplanatic design.

D. <u>EQUIPMENT ADJUSTMENT</u>: Proper adjustment is the very keynote of obtaining the best from a microscope, and is the most important activity which prevents degrading the instrument to the level of a super uncorrected magnifying glass. Resolution depends quite as much on the skill of the microscopist as upon the quality of the instrument.

    1. Illumination

        a. <u>Köhler illumination</u>: For transmitted light examination of most microscopic objects, the illumination system for obtaining the best resolution is Köhler illumination. There are other systems for more specialized conditions of observation, or for specialized objects, which are treated elsewhere.

The Köhler system requires a lamp with a small concentrated filament, a large adjustable condenser lens and an iris diaphram, close in front of the lens, which is used as the actual light source. That is to say, the image of the area surrounded by the diaphram on the illuminator (called the field diaphram) is ultimately focused into the object plane as a self-luminous source, at least insofar as the brightness and evenness of illumination are concerned. The steps in practical adjustment that are necessary to attain this condition are (refer to Figure 13):

    (1) An object slide is placed on the stage of the microscope and the object focused. This is done without particular adjustment of the light source. It is merely adjusted (if necessary) to get sufficient illumination of the object to permit this preliminary focusing.

    (2) The field condenser (on the lamp) is focused

EQUIPMENT ADJUSTMENT

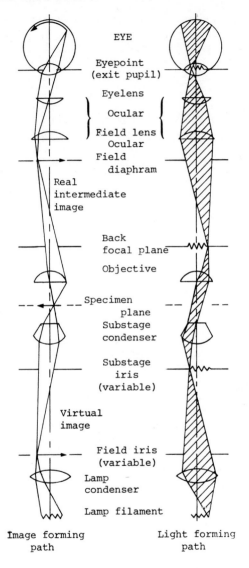

Figure 13. Köhler brightfield illumination (adapted from a Carl Zeiss illustration).

such that an image of the lamp filament appears in the plane of the substage iris. This adjustment can be made while looking at the back focal plane of the objective with the ocular removed, and the field iris wide open. Alternately, an adjustment close enough for the purpose is to focus the field condenser such that an image of the lamp filament appears on the plane surface of the substage mirror. This is aided by the use of a piece of tissue or other opaque surface being held there momentarily for the purpose.

(3) Close the field iris completely, or almost completely, and focus an image of it in the object field by racking up or down the substage condenser. If an Abbe condenser is used, the edges of the field iris will be somewhat fuzzy. With a corrected condenser the image of the iris will be sharply defined. This adjustment is made, of course, with the ocular in place.

(4) Center the focused image of the field diaphram in the field of view by tilting the mirror.

(5) Open the field iris until it just delimits the field. (At this point the area of light enclosed by the field iris is essentially a self-luminous source in the object plane.) Köhler illumination is thus obtained.

In step 2 above, it was assumed that the illumination lamp was separate from the microscope. Instruments with built-in illuminators simplify the adjustment somewhat, as the lamp is very close to the focused position and there is no substage mirror involved. The adjustment is essentially the same, however, and involves making sure the lamp iris is in focus in the object plane. Sometimes the built-in illuminator requires condensers to have an "auxiliary" lens just below them to image the field diaphragm in the specimen plane without racking the condenser down so far as to sacrifice aperture (Figure 14).

In the steps above, the substage iris, associated with the concondenser, is wide open. Its adjustment and refinements of other substage adjustments are in following paragraphs.

b. <u>Light filters</u>

(1) <u>Intensity</u>: Light intensity can be too high in many cases, wherein there is considerable scattering and therefore, due to reflections from glass surfaces (lens, slide, coverslips etc.) and internal reflections (objective barrel, microscope

EQUIPMENT ADJUSTMENT

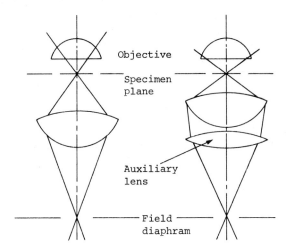

Figure 14. Köhler illumination built-in source.

tube, ocular barrel etc.), causes "glare" to such an extent as to obscure detail. Also, it might be of such intensity as not to be optimum for the eye. Filters for reducing illumination intensity are termed "neutral density" filters, and can be placed anywhere in the path of the light from the source to the substage iris. These filters reduce the intensity of the illumination in definite proportions. They can be obtained with various degrees of light attenuation and can be combined to achieve an optimum light intensity (see Table III). They do not appreciably change the spectral distribution, an advantage where color fidelity is of importance.

(2) Spectral: Objectives are corrected for spherical aberrations and chromatic aberration at certain wavelengths of light. The corrections of oculars and other optical components are corrected on the same basis. Many of the problems associated with objective spherical aberrations can be circumvented by the use of monochromatic illumination. In fact, the achromatic objective compares very favorably with the most expensive apochromatic objective (of the same NA) when used with yellow-green light.

Of course, the disadvantages of using a monochromatic source of light are obvious. Any observation having to do with the rendition and interpretation of "natural color" is thereby made impossible. Also, stained specimens, their interpretation

Table III
Combinations of four neutral density filters

Light transmission	Filter				
100%	-				
50%	0.50				
25%	0.50	+	0.50		
12%	0.12				
06%	0.12	+	0.50		
03%	0.03				
1.5%	0.03	+	0.50		
0.75%	0.03	+	0.50	+	0.50

and even the contrast quality of some images may be completely destroyed in this manner. However, where specimens can be profitably observed under light of a single color, the performance of most optical elements of the microscope can be improved and, in the case of objectives, the much more inexpensive achromats can be used to obtain resolution close to that of the apochromats.

While monochromatic light sources are available, the simplest and least expensive way to attain illumination of the proper wavelength is by the use of an appropriate filter especially an interference filter. Since the spherical aberrations of the achromatic and fluorite objectives are corrected for the yellow to yellow-green portion of the spectrum, the use of a yellow-green filter improves the spherical corrections by blocking out or attenuating the poorly corrected colors. Of course, these types of filters do not equate exactly to the use of monochromatic light, but do confine the illumination to a narrow enough band of frequencies as to be highly beneficial.

Not only does the use of a yellow-green filter improve the performance of objectives as to their residual spherical aberration, it also improves the acuity of vision by reducing the chromatic (focusing) errors of the eye. In addition, it also cuts down the intensity of the light that otherwise might have to be attenuated with additional neutral density filters and, with certain types of stained objects, actually increases contrast. If substage condensers are in use that have only been corrected for spherical aberration (the so-called aspheric or aplanatic types),

EQUIPMENT ADJUSTMENT

their performance is likewise improved to nearly that of a good achromatically corrected condenser by the use of such filters.

The use of a green or yellow-green filter with the light microscope is probably one of the easiest and least expensive ways of improving its overall performance. An appropriate combination of filters for this purpose is the Kodak Wratten numbers 15 and 58 or, better still, a 540 nm interference filter (Table IV).

For visual work, wherein color fidelity is of considerable importance, incandescent sources of light are best corrected by the use of Kodak Wratten visual M number 78AA (dominant wavelength is 473 nm).

Table IV

Important filters for improving resolution

Wratten No.	Dominant λ(nm)	Luminous trans. %	Color	Remarks
15	579	66.2	deep yellow	combine 15 and 58
58	540	23.7	tricolor green	for yellow-green
45	481	5.2	blue	for high resolution and contrast
45A	478	2.8	blue	for highest resolution

The shorter the wavelength of the light and the smaller the lambda term in Equation (1), of course, the more the resolution improves. The Kodak Wratten filter 45A was designed specifically for the highest resolving power in visual microscopy. The 45 has a somewhat narrower transmission band, but a higher transmission level than the 45A and is also excellent for improving resolution.

Filters for other purposes, improving contrast, and for photomicrographic purposes are treated elsewhere.

c. Ultraviolet: With an NA of 1.0 and a filter with a spectral transmission in the range from 610 μm to the red end of the spectrum we may resolve 80,000 lines to the inch, while with a filter having transmission in the band from 370-510 μm we can resolve 100,000 lines to the inch. Thus, with filters restricting the transmission spectra of the illumination to the blue-violet region we can definitely improve resolution.

As an ultimate extension of limiting the illumination to the shorter wavelengths, a microscope can be used with an ultraviolet light source and thereby increase resolving power still further.

With approximately twice the resolution as that attainable in the visual portion of the spectrum, the ultraviolet region becomes very attractive for that reason alone. Other ramifications in the use of UV light are involved with specimen reaction as to differential absorption etc. It is necessary at the wavelengths in the UV to employ special optics in the microscope and associated optical apparatus. A special UV condenser lens is used at the lamp illuminator, and is typically constructed of quartz and fluorite lens elements. Microscope objectives for UV use are available in both dry and water immersion types. Construction of the objectives is exemplified by those available from Bausch & Lomb. Two spherical reflecting surfaces are combined with refracting lens elements of quartz and fluorite. One refracting element serves as a support for one of the reflectors and combines with the other refracting element to complete the correction of the optical system.

Substage condensers of special construction are also necessary, and commonly a matching condenser for each different objective is employed. They are designed to work through nominal thickness (1.0 mm to 1.3 mm) of quartz microslides.

For projecting the image to recording film, special oculars are employed. Magnifications as low as 3.5X are used to provide sufficient intensity at the film plane when UV energy is at a minimum.

A special UV Image Converter is available that attaches to the ocular. It allows visual viewing of the ordinarily invisible UV image. The heart of the converter is an RCA "Ultrascope" employing a photoemissive cathode, a single stage electron imaging system and a fluorescent screen. This results in a double conversion of ultraviolet light to electrons to visible light.

Photomicrographs in the near UV can be made using a microscope with conventional optics (glass lenses). However, the lower limit is about 350 nm as optical glass will not transmit shorter wavelengths.

EQUIPMENT ADJUSTMENT

Mercury arc illuminators are commonly used in combination with filters to obtain the shorter wavelength illumination. The exact region of the UV most advantageous in producing the proper balance of resolution and contrast will certainly be dependent upon the characteristics of the specimen and how it is prepared (mounted, stained etc.).

2. Substage condenser

   a. Centering: The condenser should be centered for best results under most conditions. There are several ways of centering the condenser.

   (1) The center of the top lens can be marked with an ink dot and centering accomplished by making adjustments to bring it into alignment with the optical axis.

   (2) If the substage iris is attached to the condenser (which is often the case) the centering can be accomplished by observing the image of the closed iris with a low powered objective.

   (3) The back of the objective may be observed through a "pinhole" cap over the top of the microscope tube. A simple rimmed disc that fits snugly over the tube with an accurately centered pinhole is satisfactory.

   (4) The Ramsden disc of the ocular may be magnified with a 10X magnifying lens. The back focal plane of the objective is then visible and the alignment simply accomplished.

   (5) When available, of course, a Bertrand lens in the bodytube is an ideal way to view an enlarged image of the objective back focal plane. Would that biological microscopes possessed this accessory as do polarizing microscopes.*

In the above, it is assumed that there are centering adjustments available for the condenser. This is usually the case with the more sophisticated stands. However, with simple "sleeve" type fittings for the condenser, centering can be somewhat of a problem or not possible. Sometimes, there are screws serving to provide some adjustment even on the sleeve mounts, or paper

---

*Perhaps a better overall solution would be for biologists to use polarizing microscopes.

or metal shims can be placed between the sleeve fitting and condenser to maintain its centered position if sufficient space is available.

  b. <u>Vertical adjustment</u>: The condenser is adjusted vertically for Köhler illumination, such that an image of the field diaphram is in the object plane. As mentioned previously, the Abbe will show a rather fuzzy image in this respect (due to its inherent spherical aberration), but a corrected condenser will show a crisp image of the diaphram when it is properly adjusted. To test a condenser for pronounced spherical aberration, two steps are necessary:

  (1) The field diaphram is adjusted in size so as to give (on the slide) a focused image which is somewhat smaller than the object field.

  (2) The edges of the image of the field diaphram are observed when focused, along with the object, by different objectives at different condenser apertures (by opening and closing the substage condenser iris). The focus of the condenser (vertical adjustment) should not require much changing for the different apertures, unless there is a considerable spherical aberration present (Figure 15).

With built-in illumination stands, many of the problems of illumination and substage condenser adjustment are simplified, as there is a unified design making some of them unnecessary. Of course, this does, in some cases, contribute to a bit less flexibility for the microscopist.

With the lamp external and separate from the microscope stand, and with the use of a highly corrected condenser, some attention must be paid to whether the vertical adjustment of the condenser can accommodate various thicknesses of glass slides, and be operating at its proper position. If this is neglected, it is very possible that the advantages of the corrected condenser are being lost or derogated. To check the vertical position of the condenser the following is in order:

  (1) The back focal plane of the objective is observed, with a pinhole ocular, by Bertrand lens or by magnifying the Ramsden disc of the ocular.

  (2) The condenser vertical adjustment is changed in an up-and-down movement while the back of the objective is observed, a high powered objective being used

EQUIPMENT ADJUSTMENT 39

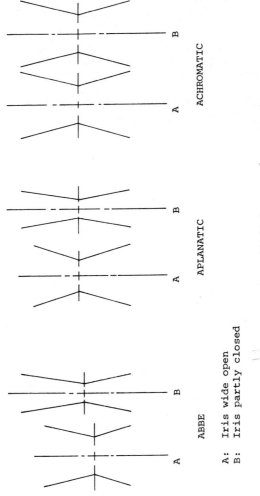

Figure 15. Cones of light from condensers.

A: Iris wide open
B: Iris partly closed

and about a nine-tenths condenser cone of light. If a marginal ring of light comes by moving the condenser up, the condenser is in an undercorrected position, insofar as the slide thickness is concerned, and if the marginal ring is found by moving it downward (less likely), it is in an overcorrected position.

Figure 16 illustrates the various conditions seen at the back focal plane of the objective. In the upper row (ABC), the slide is either too thick or the lamp is too far away. In row (DEF), it will be noted, no marginal ring appears whether the condenser is raised or lowered, and the proper conditions are obtained. In the lowest row (GHI) the slide is too thin or the lamp is too close to the condenser.

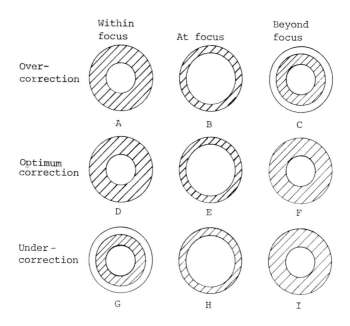

Figure 16. Condenser ring test.

Adjustments of the lamp distance are made until the conditions observed in the center row of Figure 16 are obtained. Once done, there is rarely any need for readjustment unless

EQUIPMENT ADJUSTMENT

much thinner or thicker slides are being used. If critical microscopy is being practiced, the ring test and proper adjustment of lamp distance will provide that the condenser is operating at the proper conjugate foci at which its spherical aberrations are corrected.

    c. <u>Substage iris diaphram</u>: The inner illumination disc of light (see Figure 16) as observed at the back focal plane of the objective indicates how much of the aperture of the objective is being utilized by the illuminating cone of light from the condenser. Ideally, from Equation (1) we should be using the maximum aperture available in order that the NA term be as large as possible and therefore maximize the resolution of the image. However, the term C should be as small as possible, and this can only obtain (in part) when the objective is highly corrected. Because of the design of lenses on a spherical basis, the maximum aperture of the lens does not correspond to its maximum corrected operating condition because of off-axis and edge-effect aberrations. Therefore, there is some point to which the maximum used NA of an objective can be decreased that will reduce the effect of residual aberrations to a greater degree than it will reduce the beneficial effects of the available NA. Quite often, in various handbooks there are rules-of-thumb which state that the substage condenser aperture should be set for such-and-such percentage of the total field, as observed in the back focal plane of the objective. These "rules" are to be used with caution and as guides only. When the NA of the illumination has reached the value at which the smallest resolvable detail is still reproduced with slightly increased contrast, optimum image quality prevails. This is the condition under which the microscope should be operated. With a slightly reduced NA of illumination to produce optimum image quality, there is no significant increase in the value of C in Equation (1). There is no fixed ratio to decrease the NA of the illumination source to the objective NA. This must be done on a personal basis and on the type of material being examined. Therefore, substage diaphram adjustment is a matter of experience that the observer has with respect to what he desires to see in the way of detail.

    3. <u>Coverslip thickness</u>: Spherical aberration produces a blur or fuzziness of images in brightfield, and produces halos in darkfield. This defect always results in a degradation of image detail and, if critical microscopy is being practiced, it should be eliminated or minimized wherever and whenever it occurs.

a. <u>Effect of coverslip thickness</u>: Objectives for use with covered objects are designed to be used with a specific thickness of coverslips. The two most common design thicknesses are 0.17 and 0.18 mm. From the designer's point of view the coverslip is regarded as a plano-parallel plate interposed between the objective and the object. The computed thickness of the plano-convex front lens element of an objective is reduced in design by the assumed coverslip thickness and spherical aberrations accounted for in the objective design on that basis. This is a reasonable approach since a plano-parallel plate (Figure 17)

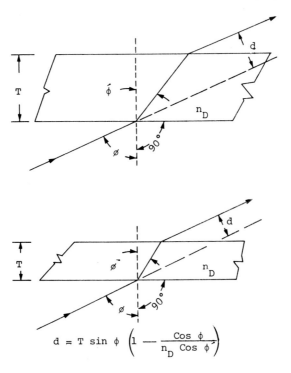

$$d = T \sin \phi \left(1 - \frac{\cos \phi}{n_D \cos \phi'}\right)$$

Figure 17. Displacement of light rays by a parallel plate.

produces the same effect wherever it might be placed across a cone of rays and, as the refractive index has only a slight effect on the aberration of a plate of given thickness, the coverslip will

EQUIPMENT ADJUSTMENT

be practically a perfect substitute for the plate of equal thickness sliced off (in design) the front lens of the objective, and the corrections of the aberrations will be correspondingly undisturbed. Many high power dry objectives are marked as to the coverslip thickness for which they are corrected. Use of the objective with its designed thickness of coverslip will minimize spherical aberrations to the designed limitations. The use of coverslips too thick or too thin will allow spherical aberration of greater than the designed minimum to occur and a consequent degradation of resolution capability.

Since objectives are designed in a spherical undercorrected condition to offset the spherical overcorrection introduced by a specific thickness of coverslip, a deviation from that thickness will produce an over- or undercorrected spherical aberration condition (Figure 18). This assumes that the specimen is in very

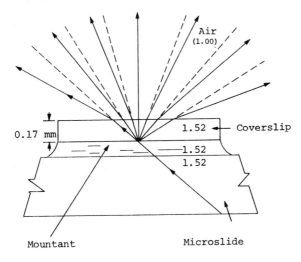

Figure 18. Equivalent spherical aberration introduced by coverslip.

close contact with the underside of the coverslip. If the coverslip deviates from the thickness required by the design of the objective, or if there is considerable gap between the specimen and the underside of the coverslip, a reduction in performance of the objective results and image quality will suffer. The magnitude of deviation from prescribed coverslip thickness that can be tolerated decreases for objectives with higher aperture (Figure 19). With the high-dry objectives of 0.85-0.95 NA the deviation

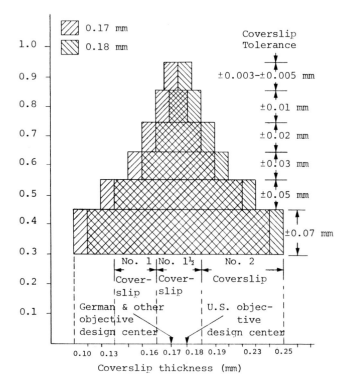

Figure 19. Coverslip thickness limits (dry objectives).

from the nominal thickness should be restricted to plus or minus 0.003 to 0.005 mm. It can be seen that if the specimen is not in close contact with the underside of the coverslip that is precisely the correct thickness, the layer of mounting medium has about the same effect as if the coverslip were too thick by a corresponding amount. Therefore, in most microscopical preparations with a coverslip there is actually an "effective coverslip thickness", composed of the actual coverslip thickness and the layer of mounting medium between the focusing plane and the underside of the coverslip. Often problems pertaining to imperfect

EQUIPMENT ADJUSTMENT

adjustment of objectives can be traced to this cause.

When the coverslip is too thick for the objective lens in use, a certain amount of overcorrection will appear in the image and vice versa. A change in thickness of 0.01 mm will produce amounts of spherical aberration at an NA of 0.65 that are large in comparison with the so-called Rayleigh limit (an optical path difference of one-quarter wavelength of the light being used).

A demonstration of the effect of coverslip thickness may be made quite easily:

(1) Focus the microscope on an uncovered object.
(2) Lay a coverslip on the object.
(3) The microscope will necessarily have to be focused upward (away from the object) to obtain an image in focus.
(4) As different thicknesses of coverslip are interposed, the focusing is changed.

Since focusing distance to the object must have changed for different thicknesses of coverslip, the degree of spherical aberration introduced is thereby changed. For the computed thickness of coverslip the aberration is minimal. For any other, it is greater, either in an under- or overcorrected amount. Generally speaking, the spherical aberration for a lens system varies with the conjugate distances of the object and image. In this latter statement lies a solution for correcting spherical aberration introduced by the use of improper coverslip thickness.

b. *Tubelength adjustment*: Microscopes which are equipped with tubes that are adjustable in length offer a method of compensating for the spherical aberrations introduced by incorrect coverslip thickness. Lengthening of the tube from the prescribed length (160 mm for instance) leads to overcorrecting for spherical aberration, and shortening to undercorrection. Adjustment of tubelength also can compensate for "effective coverslip thickness" problems wherein the object is separated from the underside of the coverslip by an appreciable amount. If the coverslip is too thick (producing spherical aberration overcorrection), the tubelength should be decreased. A change of tubelength of approximately 10 mm for each 0.01 mm departure from ideal coverslip thickness is normal.

However, even if the coverslip thickness is known exactly, and in many cases it is not, the above generalization is not a precise guide to correction. The following steps will provide a more exacting compensation for deviations in coverslip thickness.

(1) Focus on a small particle or object in the plane of the object to be examined (tubelength at normal position).

(2) If, on focusing slightly above and below this point, the appearance of the object is the same, the tubelength is proper for the coverslip thickness employed.

(3) With improper tubelength, one side of the focus will reveal a dark ring around the particle, and on the other side the particle will dissolve into a foggy image.

(4) If the dark ring persists on downwards focusing, the tubelength should be increased, as the coverslip is too thin.

(5) If the dark ring persists on upwards focusing the tubelength should be shortened, as the coverslip is too thick.

(6) After several such adjustments, it will be found that the ring becomes less and less defined on one side and begins to appear on the opposite side of the focal point until a stage is reached when dark rings, now comparatively faint, have become equally apparent on both sides of the focus. The correction is complete and spherical aberration due to coverslip thickness deviation is minimized.

The appearances in steps 1-6 above are manifestations of a crossection of the caustic being moved through the lower focal plane of the ocular as the objective is focused upwards and downwards from the "in-focus" position (see Figure 20). When the microscope tube is focused downward, shortening the object space, the caustic moves upward and vice versa.

When an object is corrected in transmitted light, as assumed above, changing to darkground illumination throws the correction off. With high-powered objectives an upward change in the drawtube of 25 mm may be required. Since darkground illumination utilizes the whole available aperture of the objective, it puts the system into a more sensitive condition and renders errors due to tubelength more obvious.

c. <u>Correction collars</u>: Microscopes of many, if not most, designs do not have any provision for tubelength adjustment. Therefore, if the coverslip is of improper thickness for the particular objective in use, some other means must be provided to minimize the spherical aberration. This is particularly

EQUIPMENT ADJUSTMENT

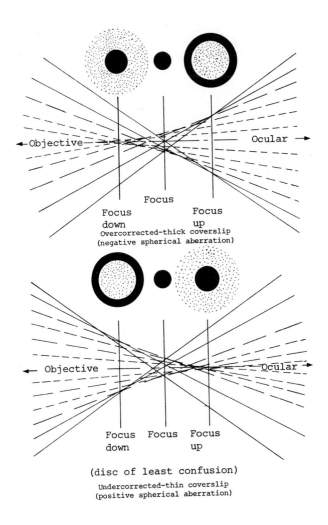

Figure 20. Coverslip effects.

true with binocular microscopes wherein mechanical variation of the tubelength is rather impractical.

In an objective, a slight change of distance between the front lens and the lens next to it will increase or decrease the free working distance and therefore the spherical aberration arising at the front plane. This allows an alternative to changing tubelength for improper coverslip thickness correction.

Basically, objectives provided with correction collars employ a convenient mechanical device whereby the distance between the front lens and remaining parts of the lens can be delicately and continuously varied (Figure 21). In a common form of

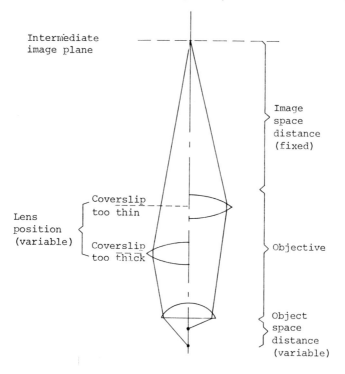

Figure 21. Correction collar action.

the device, the front lens, or sometimes the two lowest ones, are attached rigidly to the mount. The remaining inner lenses are attached to an accurately made cylindrical tube which can

EQUIPMENT ADJUSTMENT

slide with slight friction, but no lateral play, in a corresponding bore of the mount. The latter has a longitudinal slot cut through it to guide a sliding block attached to the inner tube. Screw threads cut upon the projecting face of this sliding block are engaged by spiral grooves on the inside of a collar that is rotatable on the mount, and a delicate axial movement can thus be imparted to the inner tube. One-sided spring pressure on the inner end of the latter prevents backlash. To maintain exact centering and accurate and smooth adjustment, the manufacturing costs are high for this type of objective. However, it does accomplish spherical aberration correction with far less disturbance of other objective corrections, and the magnification remains practically unchanged. For instance, the magnification with an objective of nominal 40X will only change between the limits of approximately 39.6 to 40.6X for a coverslip thickness change of from 0.14 to 0.20, respectively. With tubelength correction for similar coverslip thickness deviation, the magnification will vary from approximately 35 to 45, respectively. Also, with the correction collar, the object can be kept under direct observation while the adjustment is being made. Most of the more modern objectives with correction collars incorporate a graduated knurled ring with which the adjustment is made. Optimum image quality is obtained when the figure corresponding to the "thickness" of the coverslip used is opposite an index mark. The range of the adjustment is usually from about 0.10 to 0.25 mm coverslip thickness. This range takes into account the "effective" coverslip thickness mentioned previously wherein the object is an appreciable distance from the underside of the coverslip.

Since in most practical cases, the coverslip thickness itself, or the effective thickness, is not directly known, some indirect procedure is necessary for the determination. A method, according to Zeiss is:

> Using a 40X, 0.65 NA objective, the condenser is stopped down to half the objective aperture, and the microscope successively focused on the surface of the coverslip and the focusing plane with the aid of the fine adjustment. The corresponding readings of the fine adjustment are noted. The difference between the two settings gives the optical thickness of the coverslip, which has to be converted to the mechanical "effective thickness" by multiplying it by a factor K.

The latter is determined by means of an experiment. For this purpose the thickness D_1 of a few ordinary coverslips is determined accurately by means of a measuring aid (micrometer, dial gauge etc.), whereupon their optical thickness D_2 is determined with the microscope as described above. The desired factor K results from the two measurements:

$$K = \frac{D_1}{D_2}$$

It is not, as one might assume, identical with the refractive index (R.I.) of the coverslip. In many cases, the R. I. of the mountant may differ considerably from that of the coverslip, and if the object is of some distance below the underside of the coverslip in the mountant, this method will only be approximate. The best method is to make adjustment of the correction collar until the focused condition reveals the best image.

    d. Tubelength corrector: The tubelength can be optically, as well as mechanically, varied to accommodate for the spherical aberration introduced by improper coverslip thickness. The Tubelength Corrector devised by Jackson was marketed in England for use on binocular instruments. However, these devices are no longer commercially available, as they are difficult of universal application due to differing mechanical conditions, and most manufacturers now provide excellent correction collar equipped objectives.

    e. General comments: Tubelength adjustment is not easily reset. With a 4 mm (40X) achromat, the resetting can be accomplished, perhaps, only within about plus or minus 2 mm. Therefore, for critical work "calibrated" settings are to be avoided, the adjustment best being made under direct observation. Also, it should be remembered that for a given coverslip thickness and position of object with respect to its underside, there is a unique plane for which the correction is applied.

After correcting the objective for coverslip thickness there are two important circumstances in which the correction can and should be preserved. First, if the ocular is changed and the image is not in sharp focus (as with the initial ocular), the draw-tube or correction collar should be adjusted — not the microscope focus. Second, if photomicrography is attempted, the visual image on the camera focusing screen is not the same as that of the visual eye focusing point. Again, the microscope should

EQUIPMENT ADJUSTMENT

not be refocused to sharpen the camera image, but instead the tubelength or correction collar is adjusted for that purpose.

Even in the case of an oil immersion objective (homogeneous design) it is not quite correct to assume that coverslip thickness doesn't matter. A 1.40 NA oil immersion objective might be sensitive to variations in the thickness of the oil film — if for no other reason than the dispersion of the immersion oil might not be the same as the coverslip. In fact, it might be best if a coverslip not be used for really homogeneous mounts and best results when using this type of objective.

Coverslip thickness is especially important today wherein the principle of rigorous homogeneity has mostly been abandoned in oil immersion objective design. Therefore, it is essential that only prescribed immersion media be used with modern objectives.

Although objectives designed for homogeneous immersion may be used, ostensibly, with covered or uncovered specimens, it is different with present day immersion objectives. If immersion objectives computed for covered specimens are also employed for viewing smears, for instance, which are left uncovered as an expedient, there will be some degradation of the image. For critical work, therefore, the coverslip (of proper thickness) should be employed. It is possible, of course, to make up for the lack of coverslip by the use of immersion oil of a higher refractive index or by adjusting the tubelength etc. Therefore, it is imperative for critical microscopy, that the microscopist knows the characteristics and designed conditions under which he employs oil immersion objectives in relation to immersion media and coverslip thickness.

4. <u>Immersion oil</u>: Immersion oil contributes to a higher degree of resolution and brightness in the microscopical image. It is used with objectives whose initial magnification ranges from 40X to 120X and which are designed for immersion use. The angular aperture of objectives may vary from a few degrees for very low powers (Table I) to more than 130° for oil immersion ones. The theoretical limit for angular aperture is, of course, 180° for dry objectives but a practical limit is about 140°. This latter figure represents in air an NA of 0.95 which is the highest obtainable in a "dry" design. If the NA is to be higher, then an immersion system must be used.

Because the visible spectrum peaks at about 589 nm and that wavelength is near the optical center of the spectrum, immersion oils are compounded to have characteristics centered at that point also. Most oils are standardized at 25°, some at 20°, with some others at lower and higher temperatures for special applications. A 1°C change in temperature can effect a change in refractive index of about 0.0004. Therefore, if the microscopist is working with light wavelengths at other than the D-line (589 nm), the refractive index of the immersion oil can be brought into coincidence with that of glass by a temperature change. With known dispersion values and temperature coefficients the change can be approximated. This type of information is usually available either on the label of the immersion oil container or from tables furnished by the manufacturer. For instance, a typical set of data for Type A Cargille Immersion Oil is contained in Table V. Eleven different properties or characteristics are listed. If the light being used is in the blue region as with Wratten filter number 45A, then from data in this table the necessary temperature change to cause Type A to have an R.I. of 1.515 is easily determined.

Since the wavelength of light using the type 45A filter is 478 nm (Table IV), we wish to determine at what temperature immersion oil Type A needs to be, such that its refractive index will be 1.515. From Table V the dispersion is $n_f - n_c$ or 0.0122 at 25°C which means that between lines F and C of the spectrum, a change of 0.0122 in the refractive index takes place. In a change of wavelength from 656.3 to 486.1 nm the refractive index changes, then, at the rate of 0.0122 divided by 170.2, or 7.15×10^{-5} per nanometer. The wavelength using the 45A filter is 486.1 - 478.0 or 8.1 nm lower than the F line. Therefore, the refractive index change of Type A at 478 nm from 486.1 nm is $(8.1 \times 7.15 \times 10^{-5})$ or 57.8×10^{-5}. Thus, the refractive index of Type A at 478 nm is 1.5238 +0.000578 or 1.524378 or approximately 1.5244 at 25°C.

Now, the refractive index increases as the temperature decreases and vice versa. In this case, we wish to decrease the refractive index of Type A at 478 nm from 1.5244 to 1.515, a difference of 0.0094. This will require an increase of temperature above 25°C. From Table V the temperature coefficient for temperatures 25°C to 35°C is 0.00034 which means the refractive index changes at that rate /°C in that range of temperatures. Therefore, we will need to increase the temperature by 0.0094 divided by 0.00034 or 27.6°. That means that the oil temperature, to have the required R.I. of 1.515 at 478 nm, must be 25 + 27.6 or 52.6°C or 126.6°F.

EQUIPMENT ADJUSTMENT

TABLE V

Immersion oil characteristics

	Type A
Refractive index	
F line (4861Å)	1.5238
E line (5461Å)	1.5180
D line (5893Å)	1.5150
C line (6563Å)	1.5116
Dispersion	
$n_f - n_c$	0.0122
Abbe V	42.2
Temperature coefficients	
$\left(-\dfrac{d_n}{d_t}\right)$ 15–25 °C	0.00037
25–35 °C	0.00034
Stability ($d\,n_D$ 25 °C after 24 hrs. at temp.)	
60 °C	0
100 °C	+0.0002
Fluorescence	
(ultraviolet) shortwave	very very low
longwave	very low
Color	
(gardner)	<1
Viscosity (25 °C)	150 cs (low)
Density @ 20 °C	
GM/ML	1.03
lb/gal	8.59
Cloud point	0 °C
Flash point	
(Cleveland opencup)	410 °F
Neutralization equivalent	0.015 maximum

(data courtesy R. P. Cargille Laboratories.)

For the Wratten number 58 filter, the temperature of Type A immersion oil will have to be about 90°F for it to have an R.I. of 1.515 at 540 nm.

Another important quality of immersion oils is viscosity. Type A shows a viscosity of 150 centistokes which is very low. It would not be suitable for use where wide gaps need filling such as with long focus objectives or where the instrumentation is not vertical. For very wide gaps and for horizontal instruments, much higher viscosity immersion oils are available. Cargille Type B at 1250 centistokes (considered high viscosity) and Type VH at 46500 centistokes (considered very high viscosity) are examples.

Since the purpose of immersion oils is to provide a homogeneous path for light, any bubbles present are detrimental to the purpose, and will reduce clarity and impair resolution. In application, avoid "dabbing" the oil as it promotes bubble formation. Apply the oil to a lowered condenser, then raise it into position. Apply oil to the slide and/or coverslip, then lower the objective. Drying oils such as cedarwood oil and other natural oils should be removed after use with a suitable solvent such as xylene. Use no solvents which might attack lens cements used in optical assemblies. Even "nondrying" oils should be removed after use, as they are excellent dust accumulators.

5. <u>The use of the objective back focal plane</u>: As the various adjustments and observations pertaining to critical operation of the light microscope are made, the use of the back focal plane of the objective becomes more frequent.

The inexperienced microscopist never uses the back focal plane or even knows of its importance. The experienced practicing microscopist knows of the importance of such observations and uses them frequently in routine, and not-so-routine, adjustments of his instrument. The theoretician searching for new approaches to light microscopy or for more complete explanations of rather poorly understood light microscopy phenomena uses observations and optical considerations at that plane extensively.

There are two general types of observations of the back focal plane of the objective that deserve further mention.

a. <u>Substage lighting conditions</u>: Because of the way the substage condenser is normally adjusted, it ultimately becomes positioned such that it focuses the substage iris into the back focal plane of the objective. This is correct and proper because under normal visual observation, with the ocular in place, the

EQUIPMENT ADJUSTMENT

substage iris is not imaged in the eye, but only its effect, insofar as aperture and centricity, is experienced. Observation of the back focal plane of the objective may be accomplished easily in a number of ways. Very common practice is to remove the ocular and look down the microscope tube with the unassisted eye. A phase telescope or a Bertrand lens, when available, are excellent. Another practice, not so common, is to use a 10-15X magnifier to examine the eyepoint of the ocular while it is still in place. Any of these methods will provide a visual image in the eye of any object in the plane of the substage iris. This includes the edges of the iris itself as it is reduced to a diameter less than the effective aperture of the objective; stops and/or other devices in the filter holder just below the iris will likewise be in view.

By viewing the back focal plane of the objective, and consequently the substage iris, a determination is easily made as to the effective aperture of the objective. If no part of the iris diaphram is visible, then the objective is operating at its full NA. As the iris is closed down its edges are seen and the objective is operating at an effective aperture less than its full value by the proportional part of the light cone cut off. Conversely, the more filled with light the objective is, the closer it operates to its rated NA. The precise degree to which the substage diaphram should be closed (and consequently how filled with light the objective is) depends upon:

(1) The relative refractive indices of the mounting medium and the object.

(2) The quality of the objective (affected by residual aberrations of its design). Factor C in Equation (1) is affected by this.

(3) Coverslip thickness and any over- or undercorrected spherical aberrations.

(4) The object itself (its characteristics, such as whether it has periodic structure or is of a phase or amplitude character etc.).

(5) The characteristics of the observer's eye (factor C in Equation (1) is affected by this).

The determination of the proportion of the objective filled with light, and consequently the effective NA at which it is operating for a given object, can be very useful for reference purposes, and for calculations having to do with photomicrographic exposure.

When an objective is used at its proper working distance, as is usual for a constant tubelength and visual use (or with a fairly long bellows in photomicrographic work), the numerical aperture is proportional to the effective aperture of the back focal plane of the objective, as measured by its illuminated diameter, divided by twice the equivalent focus (Figure 22).

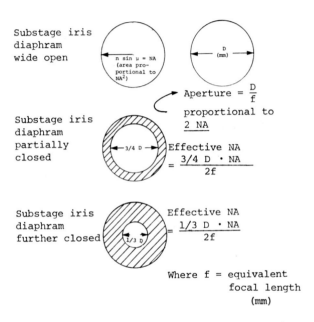

Figure 22. Views of back focal plane of objective (ocular removed).

If there are central stops, symmetrical or otherwise, in the substage filter carrier, their relative size and orientation may be observed in the back focal plane of the objective also. Again, for reference purposes and/or calculations or to afford repeatability of specific lighting conditions affected by substage parameters, this is a most advantageous observation point.

b. <u>Examination of periodic structures</u>: At the back focal plane of the objective we can see the optical phenomena produced by small periodic object structures. At any point in the

EQUIPMENT ADJUSTMENT

image plane for which the path difference to adjacent bright spots in the back focal plane (of the objective) is a full wavelength, there is a maximum of light intensity in the image plane. Regular or periodic structures, then, will show characteristic patterns of light (diffraction patterns) at the back focal plane. Diatoms quite often have the appropriate structure to produce such patterns. The conditions under which this phenomena is exhibited best is a perfectly coherent light source. As the substage diaphram is closed down, the coherence of the light source, as used by the condenser, is increased and the reproduction of the patterns in the back focal plane becomes more distinct.

Correlation of image characteristics with the diffraction patterns produced by an object is easily accomplished by fitting a pinhole ocular on one of the tubes of a binocular microscope so that the pattern can be viewed simultaneously with the object image (through the remaining ocular). The diffraction pattern is observed with the pinhole ocular at the back focal plane of the objective. Although this type of observation is uncommon, to say the least, more extensive use of it by competent microscopists with the appropriate background in physics and mathematics, might well enhance our understanding of the multitude of variables involved in microscope image reproduction and the attainment of maximum resolution in practice.

If a microscopist is familiar with the theory of image formation and the appearances at the back focal plane of the objective, observation at that point, in conjunction with the known resolution capability of the objective, can be used in the interpretation of structure, especially as to its fineness of detail. Theoretical maximum resolution capability figured in lines per inch or separation of points in micrometers can be calculated or taken from a numerical aperture table (Table I). They will provide criteria for what limits of resolution may be indicated by the back focal plane pattern. An idealized object pattern examined by a 0.90 NA objective will provide the back focal plane indications in Figure 23. It will be noted that with unidirectional axial illumination (condenser diaphram stopped down to a very small opening) the spacing of the adjacent interference maxima (M_1, M_0, M_1) appearing, are one-half the diameter of the aperture and the 0.90 NA objective under that condition resolves at its maximum of one-half of what is possible by full cone illumination (refer to B in the figure). At E the objective is able to resolve with unidirectional oblique illumination what it can with full cone illumination. For any structure to be reproduced in

METHODS IN ACHIEVING IMPROVED RESOLUTION

$m\phi$ = Zero order max.
m_1 = 1st order max.
m_2 = 2nd order max.
N.R. = Not revolved
L/in. = Lines per inch
λ = 0.527 µm

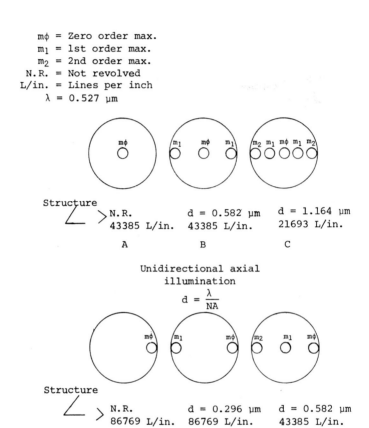

Figure 23. Back focal plane. 0.90 NA objective (maximum resolution = 86760 lines/in. or $d = 0.296\ \mu m$).

EQUIPMENT ADJUSTMENT

fidelity, there must be the dioptric or zero-order maximum plus at least one first-order maximum. An excellent basic treatment of image formation is contained in Volume 15 of this series.

c. <u>Other uses</u>: Aside from the very important adjustments pertaining to illumination, and in observing periodic structures, there are other uses to which observation of the back focal plane of the objective may be put, especially as related to the attainment of optimum resolution. Some of them have been covered previously in the adjustment of the corrected substage condenser for microslide thickness, centering of the substage condenser and adjustment for Köhler illumination. In addition, by observing the back focal plane of the objective, it can be demonstrated that a spherically uncorrected condenser is not suited to critical work. A lack of spherical correction in a substage condenser will often manifest itself when the field iris opening is small. With an objective of a fairly large aperture, it will be impossible to fill both the inner and outer zones of the objective as viewed from the back. As the condenser, with its iris open, is brought up to its position of focus, the central illuminated area of the back focal plane expands, but breaks up before the lens is entirely filled, so that finally the inner and outer zones are illuminated with an intermediate dark area, usually bluish in color. The shape of the dark area is governed by the shape of the source. This effect varies with the degree of spherical aberration present in the condenser, and produces an illumination condition which cannot take avantage of an objective's capability to resolve detail.

6. <u>The binocular microscope</u>: The foregoing discussion has assumed that either a monocular microscope was being employed and/or the difference between it and a binocular instrument were such as to be of little consequence to the discussion. However, there are differences, and perhaps it is in order here to elucidate a few facts about the use of a binocular instrument as related to resolution of the microscopical image. The following precautions are given.

(1) Unless both of the observer's eyes are known to be perfect, each tube of the binocular should be tested with the relevant eye. In making such adjustments a black card or cap over the unused tube is much better than closing the eye.

(2) To check differences in the oculars, inter-

change them. If any difference is noted, they should be appropriately marked and used in their designated location only.

(3) Registration of the two fields of view should be checked (for each ocular) to minimize eyestrain. A focused circular pattern in one ocular should show up in the other ocular in an identical position in that field.

(4) In (3) above a very objectionable prism error is indicated if the images appear relatively high or low (skewed) or divergent.

(5) Precisely adjust the interpupillary distance. This is most easily done by using low power oculars with a Ramsden disc approximately the eye pupil diameter (about 3 mm). Improper adjustment here will definitely reduce resolution of the image in the eye.

E. WORKING CONDITIONS

1. <u>Relative lighting</u>: The acuity of the human eye is found to progressively increase with illumination intensity until the intensity of sunlight itself is approached. The practice of working in dimly lighted rooms in order to "see" the microscope image better is a bad one from the standpoint of getting the most resolution at the eye. The best practice is to work in daylight near a window provided, however, that the microscope image has a corresponding brightness and color. This can be achieved by (at high powers) using a light source emitting bright light with a color temperature approaching that of daylight, neutral filters being used to control the illumination intensity.

Laboratory walls should be white and lighting good. One hundred or even 1000 foot-candles at the working surface are not excessive. The illumination should be even and free of dark or glare spots. Unless the general lighting of the laboratory is of a correspondingly high level as compared to that used in illuminating the microscope system, dazzle will be apparent and the beneficial effect of high microscope illumination nullified.

2. <u>Cleanliness</u>: The presence of dust on any optical surface of the microscope cannot in any way improve the resolution and, in all cases, will lower the ability of the optical system to deliver the resolution it is capable of. It is imperative that lens elements be free of dust, dirt and obscuring films such as oil, soot, fingerprints etc. This is true especially of the objectives and oculars of a light microscope, and perhaps to a lesser degree of

WORKING CONDITIONS									61

the condenser. The instrument, when not in use, should be covered and a cleaning routine established such that optical elements may perform to their designed limits.

Dust or oil from the eyelashes on the upper surface of the ocular eyelens is usually easily seen and removed. Dust on other lens surfaces of the ocular may be emphasized by closing the substage iris diaphram. The lens of the ocular on which the dust appears may be localized by rotation of the eyelens screw cap, and preferably should be done before any wiping takes place.

If the back lens of the objective, upon removal of the ocular, appears to be dust, it may be that the indication comes from the focus of the substage condenser. Determine which by rotation of the condenser or objective.

Moisture in very damp atmospheres can cause "fog" on lens elements internally and externally, at the same time depositing airborne dirt and/or dust particles. In humid surroundings the objectives and oculars, at least in temporary storage while not in use, can be protected with appropriate desiccants and/or plastic wrappings.

Any color or density filter used should be placed in such a location that no image of dust or marks on either of its two surfaces will appear in the field of view. For this reason they should not be put in the plane of the field diaphram, especially when Köhler illumination is being used. If they are, they must be scrupulously clean to avoid uneven illumination of the object field. A corrected condenser will show the dust particles and marks sharply in the field of view, while an Abbe will produce fuzzy darker and lighter areas of illumination.

F. OPTICAL ARTIFACTS AND SPURIOUS RESOLUTION: When the microscope is utilized in such a manner as to obtain the utmost in resolution, and high magnification is necessarily in use, the occurrence and/or presence of optical artifacts and spurious resolutions is bound to take place.

Spurious "resolution" cannot result from over-magnification as believed by some microscopists. High magnification may indeed be an important aid in recognizing optical artifacts. If extremely small linear periodic structures are to be reproduced in the image, for instance, oblique illumination is beneficial — but only if the azimuth of the illumination is correctly selected with respect to the alignment of the structures. It is these related factors of illumination adjustment (and others) that, in conjunction with high power optics, is the real basis of the belief that high power magnification results in "spurious resolution". Apparent "spurious resolution" and other optical artifacts are quite often the result of the nature of light and/or natural limitations of the optics involved. Recognition of these is very important where critical high power microscopy is being practiced. For proper interpretation of the image, the microscopist must be able to differentiate between the natural attributes of the object and artificially introduced effects either through specimen preparation or instrumentation and adjustment.

1. Optical artifacts

    a. Diffraction effects: Diffraction effects are usually apparent as diffraction fringes or complicated spurious interference patterns, or a combination of them.

More or less localized diffraction-halo phenomena may be interpreted by the microscopist as structural details of the object, such as membranes, walls or amorphous substance. These phenomena are sometimes referred to as "optical membranes".

As such, diffraction fringes can only interfere with the correct interpretation of the object image, when they are visible. It is important to know what conditions influence the relative intensities of the first and higher order maxima of the diffraction pattern if they are to be recognized for what they are, and to operate to minimize their effect on object image interpretation.

    (1) The contrast of diffraction fringes increases with an increase in coherence of illumination (stopping down the substage iris etc.).

    (2) A point-light source will increase the diffraction image visibility.

    (3) The contrast of the fringes increases with an increase of the difference between the refractive index of the object and its surrounding medium.

    (4) The contrast of the fringes is higher with monochromatic illumination than with white light.

Recognition of the fundamental features of diffraction fringes is also important. They are as follows:

    (1) The distance between the successive fringes decreases with the distance of the fringe from the fringe producing the edge.

    (2) The contrast of successive fringes decreases with the distance from the edge.

    (3) If the condenser is stopped down by closing

OPTICAL ARTIFACTS AND SPURIOUS RESOLUTION

the substage iris, the number of visible fringes and their contrast will increase, while on increasing the iris aperture, the fringes will be less visible or seem to disappear.

Since these effects do occur, some types of objects, especially those that might be expected to have a special membrane (as in bacteria, for instance), must be interpreted very carefully. Remedies for avoiding misinterpretation are:

(1) Using high magnification such that characteristics of interference fringes can be identified. Hazy border halos are seldom mistaken for membranelike structure unless insufficiently magnified.

(2) Decreasing the amount of light phase shift at the edge by changing the refractive index of the mountant.

(3) If phase contrast is used, the halos seem less obvious if negative phase contrast is used because they are dark instead of bright.

2. Other optical artifacts: Both phase and interference contrast methods, the principles of which will be covered later, very much increase the sensitivity of detecting small differences of refractive index. Of course, this is many times of distinct advantage. However, it also provides for a situation wherein optical artifacts occur more frequently.

a. Inherent to phase contrast systems

(1) Halos.
(2) Interference fringes, particularly with strongly refractive objects.
(3) Round out-of-focus images. This type of apparent structure in a phase contrast image should be distrusted until very careful focusing adjustments have made true identification possible. Images of the phase stop (the annular phase diaphram below the substage condenser) have been demonstrated to appear in a field where small transparent globular objects are. The focusing effect of such objects, acting as lenses, projects these artifacts into the image. Therefore, objects which contain within them such globular object components should be carefully interpreted in total, and not all visible features should be accepted without challenge.

b. Projection artifacts: As mentioned previously, projection of various figures, smudges, lines etc., in the plane of

64   METHODS IN ACHIEVING IMPROVED RESOLUTION

illuminating source must be reduced as far as possible or at least interpreted as to what they are. They may appear from filters etc. inserted at that point in the optical system. This condition is especially prevalent when Köhler illumination is used.

   c. Residual aberration artifacts

      (1) Spherical aberration: An increase in blur of images and also produces halos in darkfield.

      (2) Coma: This produces one-sided halos or optical membranes in the images of objects off the optical axis. Because of this asymmetry, this particular spurious effect should be easily recognizable. Coma can be introduced in a well-corrected optical system if one of the components is tilted with respect to the optical axis. In some cases this can be the object itself if it has curved surfaces that can act like lenses. The sensitivity for detecting coma effects is increased by unilateral oblique illumination.

      (3) Astigmatism: This effect may cause, in brightfield with full cone illumination, a globular small particle near the border of the field to appear as a rod.

      (4) Visual Mach effect: The human eye also plays an important part wherein optical artifacts are evident. One of the most interesting of these is the Mach effect (Figure 24). The assumption that the eye behaves as a linear detector is not true. The subjective impression of the boundary of the image is markedly different than that from the physical intensity distribution as meas-

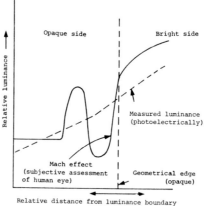

Figure 24. Mach effect.

ured by a photoelectric measuring device. The subjective effect sharpens the edge gradients in the visual assessment of luminance distribution by the human eye. This, then, may aid in detail detected by a subjective improvement in contrast. On the other hand, it may enhance optical membranes to the detriment of image interpretation. These subjective bands may be erroneously confused with and attributed solely to diffraction effects.

The effect (position, width and intensity of the bands) is variously dependent upon:

      (a) Ocular magnification
      (b) Luminance level
      (c) Wavelength of the light
      (d) Contrast of the objective relative to its background

Experimentation has shown that wide variations exist among observers in their interpretation of the visual microscope image. Great uncertainty may arise, particularly when an object image of dimensions of the same order of Mach effects are observed. The Mach effect, then, constitutes still another source of artifacts that the microscopist must be aware of if he is involved in critical light microscopy.

G. MICROSCOPE TESTING: Some comparatively simple methods by which the performance of a light microscope may be tested and of determining some of its parameters by measurement are included in this section.

    1. Parameter measurements: The dimensional aspects of any particular microscope can easily be determined with calipers, rules, micrometers and other physical measuring devices. However, there are some factors, when not supplied by the manufacturer, that entail some not so familiar methods in their determination.

        a. Objective numerical aperture: This is usually known or is indicated on the lens mount. In those cases where this is not the case, or where a confirmation is necessary, the NA can be determined. For a fairly accurate measurement an Abbe apertometer can be used. It consists of a thick glass plate in the form of a segment of a circle, and is placed on the stage of the microscope with the objective to be tested over the center of the arc of curvature. Underneath the objective the glass is bevelled so as to form a totally reflecting mirror. Light projected through the polished curved surface is reflected into the objective providing it enters the segment within the angle which corresponds to the NA of the objective. By the use of a supplemental objective screwed into the drawtube and focused on the back focal plane of the objective under test, the exact position where the marginal rays start to enter can be determined by means of movable targets (crosslines) in the segment which is engraved with the corresponding numerical aperture, and read at the position where the crossline intersection just becomes visible.

Since measurement of NA is rarely necessary, the cost of the Abbe apertometer can be prohibitive. There are other means by which NA can be determined to a fair degree of accuracy.

A very simple method is as follows:

   (1) With an objective of known NA focus on a thin-specimen slide.
   (2) Focus the light source into the specimen plane with the condenser.
   (3) Remove the ocular.
   (4) Looking at the back focal plane of the objective, adjust the condenser diaphram to just fill the back lens of the objective with light.
   (5) Measure the diameter of the diaphram opening.
   (6) Repeat steps 1, 2, 3, 4 and 5 with objectives of differing known NA's.*
   (7) Plot the diameters measured against the known NA's.

This curve can then be used to determine the NA of an unknown objective by proceeding as in the previous steps and entering the curve at the measured diaphram diameter to obtain the

---

*For best results by this method the objectives of known NA should have been measured by other precise means such as the Abbe apertometer. The nominal NA value inscribed on an objective may well be in error by 5% or even more.

MICROSCOPE TESTING

corresponding numerical aperture. If the same slide is used to obtain the data for the curve and all subsequent testing, the errors inherent in using slides of differing thickness will be obviated.

A direct measurement method is available with the use of Chesire's apertometer, consisting of a calibrated chart, a cube of wood measuring 25 mm on a side and a disc of metal with a minute hole in the center. This apertometer is very reasonable in cost (if still available) and at one time could be obtained from C. Baker Company, 244 High Holburn, London, England. Instruction for its use are included with the "instrument".

    b. <u>Eyepoint or Ramsden disc determination</u>: Focus on a thin-specimen slide preparation and then remove the slide, leaving the microscope adjustment unchanged. The intensity of the lamp should be adjusted so that a bright spot is projected on a ground glass or tissue paper surface held above the ocular. If ground glass is used, the ground side should be down. The position of the projection "screen" is adjusted from the top of the ocular until a minimum size light-spot is obtained. The diameter of this spot and its distance from the top of the eyelens can be measured with a draftsman's dividers and recorded in millimeters. In some catalogs the distance information is called "eye relief" and is noted in millimeters. It is seldom that the disc diameter is given in catalogs.

    c. <u>Objective working distance</u>: Use a thin-specimen slide and measure the distance, with the assistance of a draftsman's dividers, between the coverslip top surface and the lower surface of the objective when the specimen is in focus. The coverslip should be of the proper thickness (0.17-0.18 mm) and the specimen mounted on its underside for best accuracy (Figure 8).

    d. <u>Objective depth of field</u>: The calibrated fine adjustment of the microscope is used in this determination. Knowing the value of one division on the fine adjustment, one only needs to multiply this by the number of divisions moved between the in and out position of focus condition. Using the coarse adjustment, bring an object detail into sharp focus (focusing up). Then note the fine adjustment division. Focus down with the fine adjustment until the detail just goes out of focus and note the fine adjustment division at this point. A number of such measurements should be made and averaged (Figure 5).

e. <u>The field of view</u>: The size of the object field is easily determined by direct measurement using a stage micrometer. It is fixed for a given objective-ocular combination.

f. <u>Flatness of field</u>: Use of a stage micrometer is called for. Observation of the extremities versus the center of the scale provides the necessary information. The magnification should be selected such that the scale spans the field of view. Comparison of objectives can be made by observing how many scale intervals are in focus from center to field edge. Low power objective flatness of field is more difficult to judge by the eye, due to its accommodating ability. A photomicrograph is a better test in that case.

g. <u>Objective polarization</u>: Polarization of objectives can be checked by examining the back focal plane of the objective with both a polarizer (below the objective) and an analyzer (above the objective) in the crossed (extinction) position. Rotation of the objective should make no change in the dark "crossed polars" condition. This is easily done with a polarizing microscope or by the use of Polaroid sheet discs in a conventional instrument.

h. <u>Symmetry of illumination</u>: Observation of minute specimen bubbles is useful in this determination. In focusing upward and downward there should be no observable displacement of the image and the outer boundary of a bubble should be a perfect circle. Observation should be at the center of the field.

2. Test objects

a. <u>Thin-specimen microslides</u>: The basic requirement for object materials used in making measurements as described above is that they are of such a nature that sharp differentiation can be made of the focusing plane. Thin films, smears and various crystalline substances are in this category and are particularly suited to the task. Representative of materials for thin-specimen microslides are:

> very dilute solutions of cupric chloride, picric acid, cream of tartar and borax. Allowed to evaporate on a microslide, they provide a very thin crystalline layer. Also very small and thin inkmarks evaporated on a slide provide minute crystals. A drop of mucilage or glycerine into which air has been beaten with a glass rod will provide a "bubble" specimen. Glass micro-beads, available commercially are also suitable.

b. _For performance testing and comparison:_ The qualities of a microscope include the magnification, numerical aperture, resolution, depth of field, flatness of field, spherical and chromatic aberration corrections, and quality of assembly and execution of design. Some of these qualities can be tested and/or compared on an individual basis, others are more subjective and complex, involving both the physical and physiological. The latter tests require objects which are compatible with the many different qualities to be tested. Perhaps an ideal test object in this respect would be one that is flat, contrasty, possesses both fine and coarse structure, is best resolved in full cone axial illumination, can be examined in yellow-green light, and is mounted in a high refractive index medium. Another very important requirement of this "ideal" subject is that it be familiar to the observer. Interpretation of what is being observed plays an important role in assessing the quality of the image. Knowing what to look for, how it should look and making a comparison with that presented in the image is the fundamental process. Therefore, many test objects commonly used are of a familiar material with definite and well defined characteristics.

(1) _Abbe test plate:_ A standard 25 x 75 mm glass microslide with coverslips whose thicknesses range from 0.09 to 0.24 mm. A multiple arrangement of six coverslips or a single wedge-shape coverslip covering the same range, comprise the two styles. On the underside of the coverslip(s) is a silvered surface ruled with straight lines cutting through the film. The ruled lines are so coarsely ruled that they are easily resolved with the lowest powered objectives (Figure 25).

This plate, which requires some knowledge of optics to interpret, is used to determine optimum coverslip thickness for objectives of unknown characteristics, the extent to which the image might suffer from spherical aberration with improper coverslip thickness etc. However, other means, as discussed previously, can be used in these determinations without this elegant test slide.

(2) _Star test slide:_ This is a microslide whose surface is silvered (or aluminized) such that minute holes in the film are produced that serve as artificial "stars". The size of the holes, or stars, should not be greater than one-half the theoretical resolution of the objective

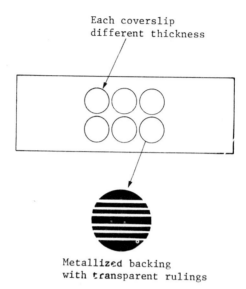

Figure 25. Abbe test plate.

Figure 26. Star test slide.

to be tested. There are various means of producing starred surfaces, including plating and vacuum deposition.

The test slide, properly prepared and appropriately suited to the objective to be tested, can be used

MICROSCOPE TESTING

to investigate spherical aberration, zonal spherical aberration, chromatic aberration, transverse chromatic aberration, coma, astigmatism, field curvature, strain, striae and fuzziness.

The star test is merely the examination of the diffraction pattern produced in different focal planes. The same procedure is used in the examination of the image of an opaque particle in making tubelength corrections for improper coverslip thickness effects. However, the diffraction disc is examined <u>critically</u> as to its structure and appearance with the star test, and <u>experienced</u> interpretation yields much more information about the optics being tested. Spherical undercorrection is more correctly termed <u>positive</u> spherical aberration wherein the marginal rays intersect the optical axis to the left (Figure 20) of the paraxial focal point, and overcorrection is termed <u>negative</u> spherical aberration because the marginal rays intersect to the right of the paraxial focal point. The significance of "positive" and "negative" is in sign convention used in geometrical optics and ray tracing. Typical light diffraction patterns on both sides of focus for these general "over" and "under" spherical aberration conditions are shown in Figure 20. When the spherical aberration is zonal (between the marginal and paraxial zones) the appearance of the patterns is similar, but with different intensity distributions among the rings of light.

With white light, chromatic aberration effects appear as colored patterns. At focus, with an achromatic or better corrected objective, a point in white light will exhibit little, if any, color. However, if the focus is varied, colors will become apparent in the refraction disc. Inside the focus (focusing downward), the disc will acquire an orange-red to red fringe. Outside the focus (focusing upwards), the disc will have a yellow-green center. The achromat, having its minimum focus at the yellow-green, exhibits this appearance, since for the types of glass used in its construction the focal distance rapidly increases for colors on either side of the minimum focus. With semi-apochromats and apochromats the combination of fluorite and glass lenses allows a "flatter" change in focal distance for other colors either side of center. Although all objectives are designed with the minimum focus in the yellow-green, the broader minimum characteristic of the semi-apochromat and the double minimum focus characteristic of the apochromat minimize traces of color either at the focus or away from it. The apochromat exhibits no noticable longitudinal (axial) chromatic aberration with this test.

Similar color changes can be noticed with very small particles on any microslide. The particle, with an achromatic objective in use, will become red on focusing downwards, and blue-green on focusing upwards. Figure 27 indicates in a rather crude fashion how color changes would occur in a star image diffraction disc upon focusing a lens not corrected achromatically. The drawing is not to scale, nor are the relative focus distances for the different colors precisely accurate.

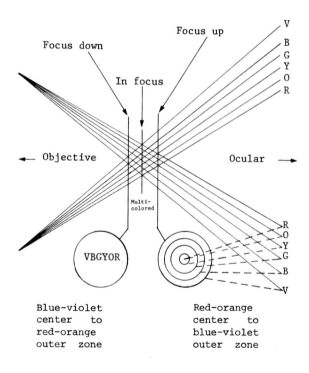

Figure 27. Chromatic aberration. Chromatically uncorrected objective.

For incidental objective comparisons, the star test slide is useful and convenient, but for authoritative analysis of objective characteristics, considerable optical background and experience in interpreting star image patterns are necessary.

(3) <u>Human blood slide</u>: A slide of human blood, stained by the Giesma method, preferably with an abundance of white blood corpuscles, approaches the "ideal" test slide. Its range and diversity of structured granulation in the nuclei and cytoplasm, vacuolation, mitotic figures etc. offer an abundance of structure capable of taxing the capacity of any objective.

(4) <u>Smear-bacterial flora</u>: A smear of bacterial flora of the teeth stained by carbol-fuschin also offers a great variety and size of objects to examine and compare. For medium and high power objectives.

(5) <u>Wing of a dragonfly</u>: For testing and/or evaluating low power objectives.

(6) <u>Insect scales</u>: Scales of butterfly and moth wings are particularly useful. For medium to high power.

(7) <u>Proboscis of a blowfly</u>: For medium and low power objectives.

(8) <u>Diatom slides</u>: Diatoms have long been used as test objects for low to high power magnification objectives. Their diverse structure and wide size range of markings offer opportunity for tests and comparisons at all magnifications. Currently available is a test slide marketed by Walter C. McCrone Associates containing five identified diatoms mounted in a row. Each diatom appears with the magnification and NA required for its resolution (Figure 28).

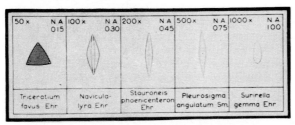

Figure 28. Diatom test slide. (illustration courtesy McCrone Associates.)

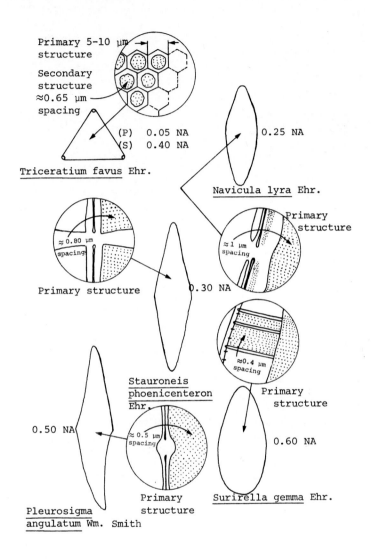

Figure 29. Test diatoms.

scope and microscopist to resolve the detail with a high-dry objective.

The value of diatoms as test objects lies in their great variation. They furnish an excellent test of the microscopist as well as the optics of the instrument. The experience gained in resolving diatom structure is valuable for a microscopist in any field.

H. <u>SUMMARY</u>: The following list summarizes the actions by the microscopist in achieving the best resolution from the light microscope.

1. Choose objectives with the highest NA and correction.
2. Use only aplanatic/achromatic substage condensers.
3. Use high magnification, corrected oculars consistent with viewing comfort and lighting conditions.
4. Select a total magnification with regard to extant conditions and not by restrictive "rules".
5. Use Köhler illumination.
6. Employ yellow-green filters, especially with achromats, whenever possible.
7. Center the entire optical system.
8. Determine that the highly corrected substage condenser is operating at its proper conjugate focus.
9. Correct for improper coverslip thickness, or use only one of prescribed thickness.
10. Use immersion oil of the proper characteristics and avoid bubble formation in application.
11. Adjust illumination for optimum used NA of an objective within its available NA.
12. Use high ambient light levels with high microscope illumination levels.
13. Maintain a high level of cleanliness of optical elements.
14. Recognize optical artifacts and, where possible, adjust conditions to minimize them.
15. Know the microscope through practice and manipulation with known test objects.

I. <u>REFERENCES AND COMMENTARY</u>: Getting the most out of the microscope in the way of resolution requires considerable background in theory and practice. The following list of references is provided to broaden the technical background of the microscopist and to present a number of different treatments and viewpoints. Appropriate comments have been added to these

REFERENCES AND COMMENTARY

entries.

1. Belling, John, The Use of the Microscope, McGraw-Hill, 1930.

A classic reference in using the microscope. The practical approach is emphasized and check-lists are used to guide proper operation and adjustment of the microscope for best performance. A bit dated in treatment, but nonetheless a very good reference.

2. Olliver, C. W., The Intelligent Use of the Microscope, Chemical Publishing Company, Inc., 1953.

Has some very handy pictorial-note charts to indicate comparative operation and adjustment.

3. Jones, F. T., "Two apparent exceptions to Abbe's theory of resolution", The Microscope, 16 4 (1968).

4. Watrasiewics, B. M., "Visual Mach effect in microscopy," The Microscope, 15 281 (1966).

5. Rowe, S. H., "The role of the Mach effect in critical microscopical investigation," The Microscope, 15 216 (1966).

6. van Duijin, Jr., C., "Visibility and resolution of microscopical detail," The Microscope, (in 13 parts) 11 196, 222 (1957); 11 254, 273, 301 (1958); 12 16, 38 (1958); 12 92, 131 (1959); 12, 185, 201, 269, 298 (1960).

This series of articles is most informative, and for the microscopist seeking to improve his knowledge of critical light microscopy a very important source. A fresh and practical approach is brought to bear on all aspects of visibility and resolution. Very highly recommended.

7. Zieler, H. W., "What resolving power formula do you use?" The Microscope, 17 249 (1969).

An excellent lucid article on the various resolving power formulas, providing a basic background on image formation and resolution. Provides recommendations as to the best resolving power formula and conditions under which it is valid.

8. Rochow, Rowe, Thomas and Kirkpatrick, "Light and electron microscopical studies of Pleurosigma angulatum for resolution and detail and quality of image," The Microscope, 15

177 (1966).

An interesting approach as to what is revealed by the microscopic image and possible ways and means of verification is presented.

9. White, G. W., "The exit pupil in visual microscopy," JQMC, 29 229 (1964).

The exit pupil is defined with a detailed account of its intimate connection with the resolving power of the microscopy.

10. "The President's address on common sense in microscopy," JQMC (Ser. 4), 4 275 (1956).

Emphasizes the role of the human eye in visual microscopy and what must be done to accommodate its functions.

11. Baker, John R. and A. Stewart Bell, "Experiments on the illumination of microscopic objects," JQMC (Ser. 4), 4 261 (1955).

12. Slater, P. N., "The star test — its interpretation and value," JQMC (Ser. 4), 4 415 (1967).

13. Hartridge, H., "The optimal conditions for visual microscopy," JQMC (Ser. 4), 4 57 (1954)

Some important observations concerning the diameter of the Ramsden disc and its relationship to the attainment of good resolution.

14. Berek, M., Optik 1, 1946; 3 4, 1948; 4 6, 1948-49; 5 1-2 1949; 5 3, 1949; 6 1, 1950; 6 4, 1950.

In this material it is shown that "critical" illumination and Köhler illumination are theoretically equivalent in resolving power.

15. Conrady, Applied Optics and Optical Design, Part I 1957 and Part II 1960, Dover Publications, Inc., New York.

These volumes are classics in practical optical design and contain some very important information for the complete understanding of why objectives are designed to perform as they do. The treatment is very lucid and even partial reading of chapters and/or paragraphs on objective and ocular design is very revealing.

REFERENCES AND COMMENTARY

16. McClure, A. E., "Glare control in the superficial illumination of mounted objects by means of the vertical illuminator," JQMC, 29 105, 147 (1963).

Provides information on some important factors to consider in working with the vertical illuminator.

17. "The President's address — spherical aberration in the microscope," JQMC (Ser. 4), 5 3-35 (1958).

The major components of the microscope are treated in relation to spherical aberration, its effects and correction. Particularly interesting is the treatment of substage condensers.

18. Martin, L. C., The Theory of the Microscope, American Elsevier Pub. Co., Inc., New York, 1966.

Most of the text material is of a theoretical nature requiring some mathematics and physics for detailed study. However, there are a number of very practical and easily understood topics of importance in obtaining the best resolving power from the microscope.

19. Sartory, P. K., "Effect of cover glass thickness on spherical corrections," The Microscope, 7 122 (1949).

A detailed and accurate account of the why and how of correcting spherical aberrations introduced by improper coverslip thickness.

20. Delly, J. G., "Light filters in visual microscopy," The Microscope, 17 193 (1967).

A very practical treatment of the use of filters in visual microscopy for improved resolution, improved contrast, reduction of glare, reduction of light intensity, reduction of heat and spectral emission control.

21. Geissinger, D., K. Sonstegard and R. Sonstegard, "Nomarski differential interference contrast — fluorescence microscopy: A new technique designed to improve resolution in fluorescent specimens," Mikroskopie, 24 321-6 (1970).

22. Bhatnagar, G. S. and R. S. Siroki, "Effect of partial coherence on the resolution of a microscope," Optica Acta, 18 547 (1971).

23. McKechnie, T. S., "The effect of condenser obstruction

on the two-point resolution of a microscope," Optica Acta, 19 729 (1972).

24. Nathan, I. F., M. I. Barnett and T. D. Turner, "Operator error in optical microscopy," Powder Tech. 5, 105 (1972).

25. Westheimer, G., "Optimal magnification in visual microscopy," J. Opt. Soc. Am. 62, 1502 (December 1973).

26. Barer, R., Lecture Notes on the Use of the Microscope, Oxford Blackwell Scientific Publication 84 (1968).

27. Cargille, John J., Immersion Oil and the Microscope, New York Microscopical Society Yearbook, 1964.
 An excellent brief treatment of the importance, characteristics and application of immersion oils may be obtained from R. P. Cargille Laboratories, Inc., Cedar Grove, NJ 07009.

28. Lambert, William E. and Milton H. Sussman, "The infinity-corrected microscope," The Microscope, 14 482 (1965).

29. White, G. W., "The correction of tubelength to compensate for coverglass thickness variations," Microscopy JQMC, 32 411 (1974).

CHAPTER 2

METHODS IN ACHIEVING AND IMPROVING CONTRAST

A. CONTRAST — FUNDAMENTAL CONSIDERATIONS: Contrast is the degree of difference in tone, brightness or color from point to point or from highlight to shadow in an object or image.

   1. Visibility: Even though the best corrected optics may be in use on a microscope to obtain the smallest resolvable detail, that detail still may not be visible microscopically because of lack of contrast. A microscope makes structure visible because of the optical characteristics inherent in a given specimen preparation which reacts with the incident illumination to modify it. The modification of the illumination may be manifested in a number of ways. The light intensity or amplitude may be modified by varying degrees of opacity, thickness etc. The color characteristics of the illumination may be altered by the color absorption properties of the specimen material, and the phase relationships of light rays with one another may be altered by the transmission characteristics of portions of the specimen. Any or all of these manifestations may be present in the final image presented to the eye. In order to discern minute detail or even larger variations in specimen structure, there must be an intensity (amplitude) or color difference, or both, of sufficient magnitude as to be detectable to the eye. Enhancement of contrast in the microscopic image can be accomplished in a number of ways, and is dependent to a great extent on the nature of the specimen and its preparation. Here we will examine some of the practical ways image contrast can be improved by modification or adjustment of the optical and lighting equipment.

   In the various methods of improving visibility through improved contrast, some assist in improving resolution, some act to decrease it. A continuous series of compromises almost always is necessary to achieve both as there is, for instance, invariably a loss of contrast accompanying an increase in numerical aperture, a criterion for improved resolving power. However, resolving power for its own sake is not the goal of the microscopist. He is necessarily interested in visible image details and interpretation of them. With that in mind he can choose various methods of improving contrast that, at the same time, result in an acceptable degree of resolution for his particular purpose. On the one hand, if a particular microscopic examination involves observation at low magnification and/or a recognition of structure whose magnitudes are well within the capability of the optics in use, contrast enhancement may be quite simple and methods used with a minimum regard to degradation of resolving power. On the other hand, at high magnifications where extreme resolution of very minute structure demands the most from the optical system, methods of improving contrast are usually complex and may be difficult to accomplish under those conditions. Each field of endeavor that requires microscopic interpretation of images has its own set of criteria. Each specimen examined has its own unique properties. In the methods of improving contrast contained in this chapter, the microscopist, his field of interest and the specimen material must be determining factors in choosing among them.

2. Intensity contrast: In order to discern the difference between details of a microscopic image that is of one single color or colorless or is being viewed in monochromatic light, the eye must be able to detect some difference in magnitude of the light intensity of amplitude. Some of the factors influencing this detecting ability need consideration.

a. Flare: This is an extraneous out-of-focus image produced by internal reflections within lens systems and their supporting mechanical structures. Reduction of this effect is usually accomplished by coating of lens surfaces, reducing (in design) the number of lens-to-air surfaces, judicious design placement of stops and blackening of all metal surfaces.

Where possible, objectives and other optical elements (oculars) should be selected that have been anti-reflection coated. The treatment briefly consists of the deposition on each glass to air surface of a film, its thickness equal to one-fourth the wavelength of the color of light for which maximum transmission is desired. The refractive index of the film is the square root of the index of refraction of the glass to which it is applied. The surface coating is about the same hardness as glass and, therefore, no special handling or cleaning is required beyond the usual careful methods used with fine optical parts. Coated lenses usually exhibit a purplish hue when viewed by reflected light.

In practice, if provisions for darkening the microscope tube are not already made, then lining the tube with "coffin paper" or velvet and the placement of a stop in the tube just below the ocular position will aid in minimizing the effect. Also, closing down the field diaphram around the object will be of assistance.

b. Glare: This effect occurs in the object space and

comes from light scattered from various glass surfaces, such as the microslide and coverslip and also from the mountant. One remedy to reduce this effect is to decrease the angle of the rays entering the objective by stopping down the substage condenser, and using the condenser without immersion (because total reflection will occur before wide angle rays have entered the slide). With an oil immersion objective and homogeneous immersion, no glare will occur. Because of the importance of contrast for actual resolution, as distinct from "resolving power", the best results in avoiding this condition will be obtained from using a corrected substage condenser.

    c. Coherence of illumination: Brightfield image contrast also depends on the degree of coherence of the illumination. Coherent light means that each and every individual ray or wave is in phase with every other (or all the light waves are "in step"). For this, a stationary "point" source of light is required. This theoretical "point" source cannot be realized experimentally, let alone practically. As the coherence of the illumination increases, contrast will improve. With a practical light microscope, as the substage iris is closed down, the coherence of the illumination increases, as the degree of "incoherence" is decreased by the smaller source presented to the condenser.

    d. Contrast sensitivity: Contrast sensitivity of the human eye is not constant but depends on actual field brightness. It is maintained at its optimal level up to a much higher brightness than is resolving power. A very rapid decrease of contrast sensitivity takes place with decreasing field brightness.

The large area surrounding the detail to be observed at the center of the field forms a so-called sensitizing field which determines the contrast sensitivity of the eye. If the illumination is decreased to a low field brightness there is a decrease of resolution (as opposed to resolving power) which is proportional to the product of the decrease of visual acuity times the decrease of contrast sensitivity of the eye. Remarks in the preceding chapter on resolution regarding the use of high light levels apply.

The area surrounding the detail should not subtend an angle at the eye smaller than about 20°. Normally used oculars (those above 5X) usually produce an angle at the eye in the range of 25 to 45°, approximately.

Resolution, as opposed to "resolving power", not only depends then upon the contrast between structure details among themselves, but also on the contrast between the whole structure and the total background area.

At low illumination levels, best contrast sensitivity (and resolution) is obtained if the structure to be resolved appears brighter than its surround — because then the brightness of the subject itself may sensitize the eye at a higher level than the surround.

In practical cases, the choice as to positive or negative phase contrast, or interference contrast, for instance, and as to which might be most effective, may well depend upon the ratio of bright and dark areas in the field and which of them are to be resolved.

In darkground illumination, if the area of the object is small compared to the field, the contrast sensitivity of the eye is decreased. In that case, increased contrast sensitivity can be gained by using an imperfect darkfield (background not absolutely black). This is accomplished, of course, at the expense of decreased image contrast. That choice will be dependent upon what information the microscopist is seeking from the specimen. A better solution would be to increase the brightness of the object by using a higher illumination level, so that the apparent self-luminosity level of the object is higher, and thus in this way improve the contrast sensitivity of the eye.

3. Color contrast: Visibility and/or resolution (as opposed to resolving power) at the limit of resolving power of a lens system is poor unless contrast is obtained by color differences. The color contrast may be obtained in a number of ways. Natural selective absorption by various components of the object illuminated by white light, and/or by differential staining of the object are commonly encountered conditions of observation.

Perception of color depends on the qualities of the eye, illumination level and the size of the colored details. At or near the limit of resolving power, if the smallest details are of different colors, we must be aware of some shortcomings of the eye. Because of chromatic aberrations of the eye, the planes in which details of a different color may appear may not be true (this is dependent, of course, upon the colors involved and the chromatic magnification of the eye). Also, very small colorless detail may appear colored, and in the case of very small details their natural colors may not be perceived unless they are blue-green or red. For correct color appreciation, colored objects should subtend an angle of at least 2° at the eye.

The color sensitivity of the eye depends exclusively on the retinal cones and, although a physiological treatment of the human eye is certainly beyond the scope of this book, it is informative to have some general indication of how that sensitivity is distributed across the visible spectrum (refer to Figure 30).

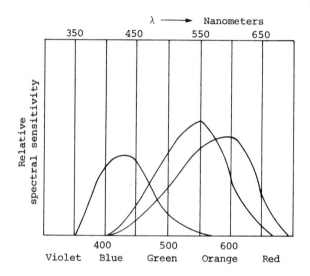

Figure 30. Relative spectral sensitivities of the three color sensitive elements of the eye (cones).

For instance, the optimum spectral sensitivity of the retinal cones is at a wavelength of about 530 nanometers (0.530 micrometers), and the optimum sensitivity of the light adapted eye is at about 588 nanometers. The quality of the color impression changes with intensity. At low intensities, the sensitivity shifts towards shorter wavelengths (yellow and red appear darker, and green and blue appear brighter).

For the reasons cited in the preceding paragraphs, it is apparent that the observation of very small detail will be inherently bound with some incorrectness of color interpretation. This can be especially important in application of cytochemical color tests, for instance, as some blue colors will often definitely give a green color impression.

The color sensitivity of the eye is influenced by the prevalent color in the field just as its amplitude contrast sensitivity is affected by the light intensity level of the field. With small details as encountered in many microscopical examinations, flare in the optical system may cause the background color to flood or wash out the detail color and thus impair the color contrast sensitivity.

The number of color differences that can be distinguished in a continuous spectrum by the eye is more than 150. Thus, for contrast improvement purposes, there is a large number of remedial measures involving color available.

B. CONTROL OF THE SUBSTAGE DIAPHRAM

1. Iris diaphram: As mentioned previously, increasing the coherence of the light will improve image contrast. Most microscopists are aware of the increase of contrast obtained by decreasing the aperture of the substage diaphram. Closing down the iris diaphram to some degree below the maximum objective diameter to offset the undesirable edge effect aberrations inherent in all spherically designed lenses has been discussed in a previous chapter. That action previously described was primarily taken to improve resolving power as expressed in Equation (1). To improve resolution it is desirable, and in many cases necessary, to further decrease the iris opening to improve contrast. This effectively increases the coherence of the light source by decreasing its size. But it does so at the expense of resolving power — the ability to separate detail. Practically, a further decrease of condenser aperture beyond a certain point by the iris diaphram does not increase image contrast further, and affects resolving power and resolution (visibility etc.) adversely. With completely incoherent full cone illumination, for instance, the limit of resolution is twice that with completely coherent axial illumination, but the contrast is much less. Therefore, as in many adjustments of the microscope, the iris diaphram opening must be controlled carefully if the maximum benefit is to be derived in interpretation of the image at the eye.

Generally speaking, in brightfield work, the substage iris diaphram should never be opened more than will admit light of the aperture of the objective in use and, as has been discussed before, usually better results can be obtained by stopping down farther than this to both minimize the objective edge effect aberrations and for increased source coherence.

If the iris diaphram is left wide open, for instance, and the condenser aperture is greater than that of the objective in use, a

CONTROL OF THE SUBSTAGE DIAPHRAM

pure darkfield effect with no component brightfield usefulness will be created. This situation should definitely not be obtained in brightfield because the darkfield effect produced induces self-luminosity to the object, thus decreasing contrast and contrast sensitivity, and increasing scattering illumination detrimental to the image.

2. <u>Oblique illumination</u>: In the preceding section the discussion has been limited to illumination that is central and aligned on the optical axis of the microscope. With some condenser-iris diaphram mounts the iris diaphram can be de-centered and thereby effect an oblique lighting condition. If this is not possible, it is possible to, in most cases, leave the iris diaphram wide open and insert fixed-opening diaphrams in the filter carrier of the instrument (in the case where it is situated directly below the iris diaphram). These fixed opening diaphrams may be devised to provide an axial or oblique light condition.

Oblique coherent illumination can increase the limit of resolution over that of axial coherent illumination. Annular illumination (a special condition of oblique illumination) will improve contrast and visibility of fine detail without affecting resolution. This obtains provided the peripheral parts of the optical system (especially the objective) are well corrected for aberrations. Even though this condition is true for fine detail, the image quality for coarse structure may be impaired.

Again, dependent upon the information to be obtained from the object, the microscopist is faced with a choice as to the desirability of such remedies and the degree to which he wishes to apply them.

C. DARKFIELD ILLUMINATION

1. <u>Darkfield stops</u>: Darkfield illumination is effective in obtaining increased contrast and for observing the colors of structures. It is useful especially in two instances: for contrast, where the specimen is colorless and the differences in refractive index between it and the mountant are low (as is quite often the case with living specimens), and where the specimen is composed of very minute particles (even somewhat below ordinary microscope resolution). The latter is called ultramicroscopy and will be briefly covered in this section.

Essentially, the object is made to appear self-luminous in darkfield, a high contrast condition. Simple darkfield can be obtained by blocking all central rays of light from the condenser from entering the objective. This is easily accomplished by the use of opaque stops in the filter carrier of the substage assembly. The only light (ideally) which enters the objective, under this condition, is that which is scattered by the specimen because it is illuminated by a hollow cone of light of too large an angle to permit the direct beam to enter the objective. This is, then, the extreme case of annular illumination mentioned in a previous paragraph. The size of stop required varies with the numerical aperture of the objective in use, the larger that being, the larger the stop required. For providing light that is restricted from entering the objective and yet of such angularity as to provide a high proportion of light scattered from the object, the diameter of such stops should be about ten percent greater than an opening of the substage iris diaphram which corresponds to full aperture illumination of the objective in use. This can easily be determined by adjusting the iris and observing the back focal plane of the objective. The measured iris opening is increased by 10 percent and the stop diameter selected to equal that figure. Stops are commercially available or can easily be made.

An improvement in this particular method of obtaining the darkfield condition is to oil the condenser to the microslide. This will provide a somewhat "blacker" field.

At best, however, this simple method is useful at only low and medium magnifications. Due to internal reflections in the ordinary condenser the best contrast at high magnifications cannot be obtained, and resolution is less than can be obtained in brightfield.

   a. Rheinberg illumination: A special condition of darkfield illumination by stop insertion in the filter carrier of the condenser is Rheinberg illumination. The central stop and/or its annular ring, in this case, are made of colored transparent or transluscent material. Thus, the background will appear in the color of the central stop (instead of being dark in the case of an opaque stop) and the object will appear in the color of the light coming through the annular portion. Many interesting and effective combinations and patterns of contrast can be obtained at low and medium magnifications by this method. A more complete treatment of special apparatus for this purpose is included in Volume 16 of this series.

   2. Darkfield condensers: For critical darkfield work, specialized darkfield condensers are used. They are of special construction and are available for both transmitted and incident

DARKFIELD ILLUMINATION

light illumination. Some manufacturers make phase contrast condensers which provide a hollow cone suitable for semi-darkfield and which may be used effectively in providing image contrast when used in conjunction with ordinary (nonphase contrast) objectives.

With darkfield condensers designed and constructed especially for that service, the resolving power at high magnifications obtained can be as high as that obtained by brightfield methods.

Two types of darkfield condensers are generally available, the paraboloid and cardiod designs (Figure 31). They are both reflecting type condensers in that the optical paths of light through them are formulated via reflecting mirror surfaces rather than refracting lenses. This provides for absence of chromatic aberrations over the spectrum and excellent spherical correction. The paraboloid is the older of the designs and can be obtained with a numerical aperture ranging from 1.2 to 1.4. It is primarily designed to be used with high power oil immersion objectives and a brilliant illuminant.

The more recently developed cardiod condenser is of a double-reflecting arrangement, has a more concentrated focus and is somewhat more difficult to adjust than the paraboloid. Typically its range of NA is from 1.15 to 1.4. It is used extensively in colloid research.

The construction of the cardiod condenser is such that the incident parallel rays are subjected to successive reflections from spherical mirror surfaces which produce a cone of light whose rays are brought to a mathematically correct focus. This optical property will produce an exact image of a very small surface in its conjugate focal plane, thereby making it useful even for ultramicroscopy.

Zeiss darkfield condensers are designated to indicate the interior and exterior limits of the illuminating cone aperture, such that the condenser used can be selected to produce the best results with objectives of a particular NA. For instance, the commonly available designations are for cone angles of:

    0.7/0.85 for objectives of NA 0.4 - 0.6
    0.8/0.95 for objectives of NA 0.6 - 0.75
    1.2/1.4 for objectives of NA 0.75 - 1.0

Zeiss designs include separate condensers for dry darkfield, and oil immersion darkfield work.

    3. <u>Practical matters</u>: Often, the cardiod condenser is cal-

culated for the lamp condenser (which is an imaged source) to be about 12 in. distant. The light source should be focused at a considerably greater distance than this. The illumination beam

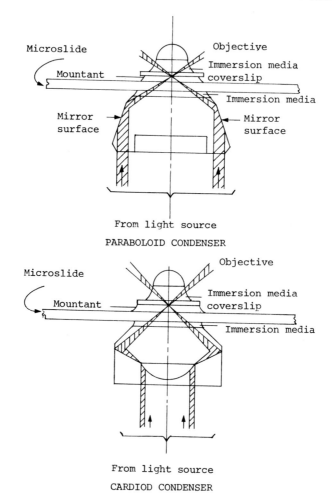

Figure 31. Darkfield condensers.

DARKFIELD ILLUMINATION

of light must cover the aperture of the condenser. It is preferable if the rest of the system is centered and adjusted first with a brightfield condenser. Then, if it is possible, substitute the cardiod and view the specimen under a lower power than will ultimately be used. If the condenser is out of focus, an illumination ring will be observed instead of a small bright disc (refer to condenser adjustment procedured in the preceding chapter). The disc must be centrally located in the field. Since the numerical aperture is high, immersion media between the darkfield condenser and microslide is always used. Cedarwood oil has been a common medium in the past, although synthetics are available now that have improved characteristics (nondrying, nonfluorescent, colorless etc.). The object should be mounted in a medium and provided with a coverslip. Oil immersion objectives with apertures (NA) greater than 1.0 must have that aperture reduced in some way. Conventional objectives can be equipped with "funnel stops" inserted at the back, or specialized objectives provided with built-in iris diaphrams can be used.

The tip of the focus of the hollow cone from a modern darkfield condenser is very fine, being formed from a very high angular aperture so that the position (focus) is sensitive to the location of the object plane, which is determined by the thickness of the microslide. The proper thickness of the microslide for a particular darkfield condenser is usually stated (within limits) by the manufacturer of the condenser and for a paraboloid, for instance, might range from 1.35 to 1.6 mm. For best results, the slide thickness specified must be adhered to.

Intense illumination for darkfield work is usually furnished by the use of carbon arcs, ribbon filamented, or compact coil filament, high intensity lamps. Since centration of the illuminant is very important in darkfield work, some manufacturers provide special darkfield illuminators. These illuminants are integrated with the darkfield condenser such that the lamp can be screwed, or otherwise fastened, into the adjustable centering mount of the condenser. Other facilities for alignment and compensation for varying thickness of slides is allowed by substage focusing mounts.

4. <u>Incident illumination</u>: Thus far, the discussion of darkfield illumination has been concerned with transmitted illumination via the substage and for subject matter suited to that method.

In reflected light microscopy, unprepared and diffusely reflecting surfaces (offering little in contrast) can be successfully examined in darkfield.

A modern design by Nikon for incident darkfield is illustrated in Figure 32. The illuminator's light beam is reflected by the

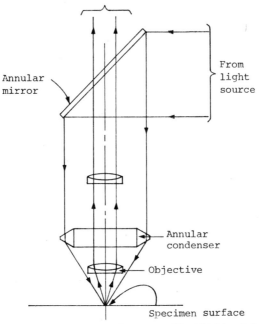

Figure 32. Darkfield-condenser incident light. (adapted from illustration courtesy Nikon.)

annular mirror through the annular condenser surrounding the objective and obliquely illuminates the object. Surface details and colors stand out in bright relief against a dark background. The image passes through the clear center of the annular mirror to the oculars.

5. <u>Ultramicroscopy</u>: Although not used as extensively as in the past, ultramicroscopy is still a very useful method of examining exudates from lesions, milk, saliva, blood, chyle, lymph, pleural, pericardal and peritoneal fluids, and petrological preparations in fine suspension. Also it is used in the study of mother liquors, fermentation of wines, vinegar and molasses. Biological applications include the study of flagellated forms in

DARKFIELD ILLUMINATION

bacteriology, and the cilia and flagellata of protozoans. Other industrial uses include the examination of flocculents, paint and paint pigments, edible and lubricating oils, and many applications in colloid chemistry.

The underlying principal of ultramicroscopy is that bodies (particles etc.) become self-luminous when illuminated by a very intense beam of light, by scattering. If enough of the scattered or deflected light is picked up by the eye or instrument, the particles will appear as small discs of light, provided they are separated by distances resolvable by the microscope optics.

The cardiod darkfield condenser is well suited to many applications in ultramicroscopy and unless very exacting particle size determinations are required, more elaborate outfits, as exemplified by the slit ultramicroscope, are not required. The latter embodies a special high intensity arc-lamp which directs a minute beam of light at 90° to the optical axis through a special condensing system using a microscope objective. The arrangement utilizes a special optical bench alignment apparatus, and an accurately adjustable slit through which the dimension of the light beam is controlled.

The detecting power of the ultramicroscope is independent of its resolving power and the wavelength of light used. It is dependent only upon the intensity of the incident light and the contrast of the field which in turn increases with the difference between the refractive index of the particle and its background. Since the wavelength of the light is not a factor, particles can be detected visibly that are below the resolving limit in size. However, they will only appear as points of light and nothing can be ascertained directly about their size, form or color. Indirect quantitative facts may be learned from observations of the Brownian movement in a liquid of known properties.

Ultramicroscopy is, then, the ultimate in obtaining contrast, as all else has been sacrificed in making particles visible, even beyond the resolving limit.

D. USE OF FILTERS: Contrast in the image of a colored microscopical preparation can be improved by the use of filters controlling the color or nature of the light used for illumination of the specimen.

1. Selective filters

   a. Colored filters: This type of filter is used to control the color of the light illuminating the object, and does so by its own color absorption and transmission properties. Common-

The diatoms are:
>   Triceratium favus Ehr.
>   Navicula lyra Ehr.
>   Stauroneis phoenicenteron Ehr.
>   Pleurosigma angulatum Wm. Smith
>   Surirella gemma Ehr.

This slide is very useful, but as with many test objects usually prescribed, what is to be "resolved" by the microscopist is not made clear. Triceratium favus Ehr. has primary and secondary structure, and Surirella gemma Ehr. has structure both longitudinally and transversely that differs somewhat. In this area of test specimens as any other, there is no substitute for intimate knowledge of the subject matter. For instance, a microscopist unfamiliar with diatoms might conclude he has "resolved" Surirella gemma Ehr. when he has not at all, and may be unaware of finer detail to be resolved. Also, although many diatoms exhibit regular structure, there is considerable variation in that structure not widely known among nonspecialists. Pleurosigma angulatum Ehr. is an exception, being remarkably constant in structure, making it especially valuable as a test object. Diatoms, especially as test objects, should be mounted in a high refractive index media (1.65-1.70) as many of them cannot be resolved in lower R.I. mounts because of insufficient contrast.

Figure 29 indicates the type of structure that is to be resolved with the diatom test slide of Figure 28. It will be noted that Triceratium favus Ehr. has both primary and secondary structure. The primary structure is that which on the test slide calls for an NA of 0.15. Beside each of the diatoms illustrated in Figure 29 is a value of NA that is sufficient if blue light with Wratten filter 45A is used with an objective that is of good enough quality that value C in Equation (1) becomes nearly as low as 0.4 (see Table I). Surirella gemma Ehr. is not too difficult to resolve if only horizontal "striations" are obtained, but to resolve the individual puncta takes much more skill on the part of the microscopist.

A diatom that has been a classic test specimen is Amphipleura pellucida Kutz. The detail varies from 0.276-0.268 µm in the form of puncta ("beads") that are quite invisible at lower powers. It is a test of micro-

ly, this type of filter is either of colored glass or colored gelatine unprotected or protected by clear glass in a sandwiched construction. Basically, these selective filters are designed to transmit predetermined wavelengths from a light source of mixed wavelengths (usually white light). Their action is to suppress undesirable regions of the spectrum by absorption of light of all other wavelengths than those desired (Figure 33). Common colors available in solid glass filters are blue, yellow, green, orange and red.

The characteristics of human vision do not enter into the determination of spectrophotometric curves; therefore, it is to be considered as a purely physical measurement. Color designates only the appearance of the light as seen by a human observer and in general has little, if any, relationship to spectral energy distribution.

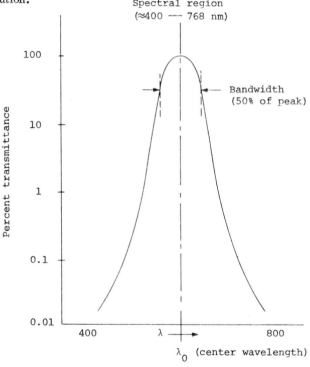

Wavelength in millimicrons

Figure 33. Visual bandpass filter characteristics.

To account for this, filters are often listed with a "dominant wavelength". This is a psychophysical counterpart of "hue", relating human perceptions to their physical causes. The dominant wavelength is defined as "the wavelength corresponding to the intersection of the spectrum locus with a straight line drawn from a point representing the light source through a point representing light reflected from (or transmitted by) a sample, on the American Standard Chromaticity Diagram." The light source is, in most cases, Standard Source "C" of the Commission Internationale de l' E'clairage (CIE), being artificial daylight of 6740° color temperature.

Dyes used in this type of filter are only moderately stable and may with time change color. Prolonged exposures of filters to daylight, particularly direct sunlight, is to be avoided because it, or exposures to high temperature and high relative humidity, may cause physical damage.

b. <u>Interference filters</u>: These filters use the principle of optical interference to accomplish selective or colored transmission.

Ordinarily the interference filter is composed of two semitransparent silver layers separated by a thin layer of transparent material (magnesium fluoride generally). The thickness of that layer is so controlled in design that multiple reflections between the silver layers are in a state of constructive interference in the transmitted beam for some chosen wavelength of light. The thickness chosen is slightly too great for constructive interference of shorter wavelengths, and slightly too thin for longer wavelengths. Only a narrow band of frequencies is transmitted.

Because the transmission band can be very accurately controlled in manufacture, these filters are more exacting in their application, singly or in multiple, to produce controlled colored light. The bandpass characteristics of these filters are also narrower than those of absorption filters. Two bandpasses are available, classed as wideband and narrow bandpass filters. In the green light portion of the spectrum, for instance (bandwidth of approximately 500 to 575 nanometers), a wide bandpass filter for a maximum wavelength of 546 nanometers may pass light of plus or minus 5 nanometers, while a narrower bandpass filter restricts it to only about plus or minus 2 nanometers (Figure 34)! Since many absorption stains used in microscopy have a fairly sharp characteristic, it can be appreciated that the use of interference type filters has some definite advantages in obtaining

96   METHODS IN ACHIEVING AND IMPROVING CONTRAST

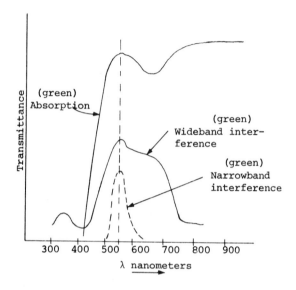

Figure 34. Characteristics. Absorption and interference filters.

maximum contrast.

In using this type of filter, it is important that care is exercised to be sure that the mirror surface is turned toward the light source. The filter should not be exposed to temperatures higher than about 66°C (150°F). Any mechanical stresses such as impacts, compressions or tensions can damage the filter. It should be stored in dry air at temperatures not higher than about 40°C (104°F).

A comparison of the transmission characteristics of absorption and interference type filters is illustrated in Figure 34.

c. Application for contrast

(1) A general guide: In selecting contrast filters, choose a filter complementary in color to the object color or stain. The best method of determining the contrast required by an object is to examine it visually with the microscope, first by means of a combination of filters transmitting as completely as possible light of the wavelength absorbed by the preparation, and then by other filters transmitting light less completely absorbed

USE OF FILTERS

until the degree of contrast obtained is satisfactory to the eye. As a general guide:

Use for <u>blue</u> colored preparations a <u>red</u>, yellow or orange filter
" " <u>green</u> " " <u>red</u> "
" " <u>red</u> " " <u>green</u> "
" " <u>yellow</u> " " <u>blue</u> "
" " <u>brown</u> " " <u>blue</u> "
" " <u>purple</u> " " <u>green</u> "
" " <u>violet</u> " " <u>yellow</u> or green "

For instance, a blue filter will materially darken a yellowish object or, if the object is blue, a yellow or red filter will make it stand out. Because several colors are usually involved in a specimen, either natural or stained, the selection by trial is almost mandatory as each combination and contrast requirement is unique.

(2) <u>Specific applications</u>: Eastman Kodak, Zeiss, Ilford and other major manufacturers provide filters, or sets of filters, with definite characteristics that may be used in specific applications or combined for special transmission or absorption applications.

The Wratten series of filters by Kodak (Table VI) are designated by number or number-letter combinations for specific spectral transmission characteristics. Zeiss publishes transmission limitations in the form of tables of figures and characteristic curves showing the spectral transmission of their filters.

The spectral limitations are usually stated to indicate what band of frequencies or wavelengths of light are transmitted through the filter from incident white light. The figures for wavelengths are usually expressed in millimicrons or nanometers (10^{-9} meters), the latter expression being somewhat more modern usage.

Common stains used in microscopy have definite spectral absorption properties. For instance, Eosin Y has an absorption band from 490 nanometers to 530 nanometers (nm). Best contrast conditions for this stain can be obtained by selecting a filter, or combinations of filters, whose transmission characteristics are in this range. This follows since for a color to be rendered as dark as possible, it must be viewed by light which is completely absorbed by the color — that is, by light of wavelengths comprised within its absorption band.

Since absorption bands produced by stains (as Eosin Y above) are fairly sharp, if the absorption band is matched as closely as

TABLE VI
Kodak Wratten filters — useful in visual work

No.	Dom. λ (nm)	Lum. trans %	Color	Remarks
15	579	66.2	Deep yellow	Combine for contrast blue preps. — also improves detail in insect mounts.
22	595	35.8	Yellow orange	
22	595	35.8	"	Increase contrast w/blue preps.
25	615	14	Red	Increase contrast w/preps stained w/methylene blue
35		0.45	Violet	Contrast
38A	479	17.3	Blue	Red absorption — increase contrast w/faint yellow or orange
56	552	52.8	Very light green	
57A	534	37.3	Light green	
58	540	23.7	Green	Contrast for faint red or pink preps.
61	537	16.8	Green	
66	512	58.3	Light green	Contrast
78AA	473	15.8	Bluish	For color fidelity w/incandescents.

possible by the filter transmission band, the greatest contrast will be experienced. However, the spectral absorption of a stained object might be different from that of the stain in a watery or alcoholic solution. Therefore, any matching of filter spectral transmission characteristics to stain absorption characteristics can serve as a guide or starting point only.

In the case of Eosin Y, Kodak filter number 58 with a transmission characteristic of 480 to 620 nm (green) can be combined with number 45 with a transmission characteristic of 430 to 540 nm, and thus provide maximum contrast. It will be noted that

USE OF FILTERS

only wavelengths which are mutually within the passbands of both of the filters will be transmitted through the pair (in this case the lower limit of one and the upper limit of the other). Figure 35 illustrates the case.

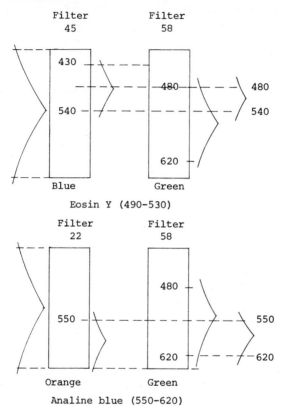

Figure 35. Kodak Wratten filter combinations for biological stains.

To improve or maximize contrast, use of the manufacturer's data (transmission band wavelengths etc.) and listings of the spectral absorption bands of the commonly used biological stains can be of considerable value as pointed out above. However, to save time and for many applications, the filters usually most effective have already been worked out and are available in convenient sets. These sets are commonly provided in 33 mm diameter circles with glass used that is of the same quality as that of coverslip in that it is extremely flat and free of variations.

In the Wratten series by Kodak there are a number of filters which are recommended for contrast improvement in visual microscopy. They include number 38A (blue) for apparent contrast with faint yellow or orange specimens, number 66 (light green) and 58 (green) for apparent contrast with pink and red specimens, stained with methyl green, methylene blue etc.

2. Liquid filters: In addition to the colored glass/gelatine and interference filters, liquid filters may be preferred for a variety of reasons.

Liquid filters are chemical mixtures or compounds contained in glass cells with plane-parallel sides. They yield excellent results and have the advantage of considerable flexibility, under the control of the microscopist at all times. In addition, cells of considerable thickness (3 cm, for instance) containing the filter liquids can also serve as heat absorbers when high intensity lighting is used. The cells designed for the purpose are about 6 cm by 8 cm by 1 cm thick. They are used when it is required that they have a minimal effect on illumination and optical adjustments. Under less stringent lighting conditions, other types of containers for liquid filters may be used, even common laboratory flasks.

The color of the particular mixture is not always a direct indication of how light is affected by it, some appearing to the eye to be of no well defined color. Examples of simple chemical filters are a saturated cupric acetate solution which serves as an effective blue-green filter, and a solution of sodium or potassium dichromate where yellow light is desired. Other liquid filters require more than one chemical, sometimes several, compounded in exact proportions. Chemical and laboratory handbooks provide directions for making up such solutions.

3. Infrared filters: In some types of specimens (for example, chitin of insects) contrast and detail can be improved by the use of infrared light. Ordinary illumination provides infrared rays, and a proper filter will transmit only the infrared portion of the emitted spectrum. A deep red filter can often be used to advantage in such circumstances.

Because of the color limitations of the eye, this method is most applicable to recording an improved contrast image on photographic film. It is, of course, necessary that focusing be done very carefully as infrared illumination at 820 nm will focus at a

USE OF FILTERS

different plane than yellow-green light, for instance. Special photographic plates (films) and filters are available commercially for the purpose. The techniques are covered elsewhere in this series.

4. <u>Polarizing filters</u>: Polarizing filters change the nature of the light impinging on and/or emerging from the subject material. Dependent upon the properties of the specimen material examined, considerable color contrast can be obtained. Many objects possess some degree of birefringence. A typical case is vegetable fibers (cotton, linen etc.) which when unstained in balsam are almost invisible, but very contrasty in color with polarized light. Insect chitin also quite often reacts to polarized light to create an improved contrast effect in the image presented to the eye.

Polarizing filters produce and analyze plane polarized light. Referring to Figure 36, light which is vibrating in one

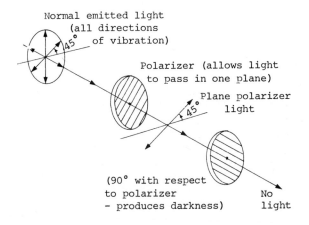

Figure 36. Action of polarized light filters.

plane only is called polarized light. Normal emitted light (a mixture of light vibrating in all directions) may be polarized by reflection, double refraction, selective absorption or scattering. Polarization enables the distinguishing of changes in structure and the composition of material that is not discernable with ordinary light. Changes in appearance under polarized light serve as an identification of many materials. The use of polarized light for purposes of identification and structure is a special topic of other books in this series. In this chapter we are concerned entirely with the effects of polarization which produce contrast improvements in the image.

The polarizer is usually placed in the filter carrier of the substage condenser such that the light incident upon the subject material is plane polarized. The analyzer is placed somewhere above the object, usually in the tube of the microscope, or just below or above the ocular. With birefringent materials, the position of "crossed polars" will produce colors that are of considerable assistance in obtaining contrasting differences between structural elements of the material.

When a material is not birefringent and, therefore, does not react with a polarizer and analyzer to produce contrasting colors, it is sometimes of real advantage to use only the analyzer. Some reflected light (from glass surfaces, specimen surfaces etc.) appearing in the image to the eye is plane polarized by virtue of its production (reflection, scattering etc.) and, if the analyzer is oriented properly, that light appearing as glare in the image can be suppressed or eliminated, thus improving the contrast in the final image.

E. <u>PHASE CONTRAST</u>: Contrast in the microscopic image can be improved by taking advantage of the optical path (refractive index times thickness) differences within the specimen which result in the existence of phase differences between the light waves transmitted by the various portions of the specimen.

Unstained, transparent specimen materials, which have relatively little variation in opacity or color to provide sufficient amplitude (intensity) variations for detail discernability to the eye, are seen in phase contrast with a detail and contrast not available with standard amplitude illuminations methods. Thus, phase contrast is useful in eliminating the need for and time to stain many materials before observing them microscopically. A great advantage is that much living material can be examined in its natural state without being killed and stained, and its behavior and ongoing living processes observed with high contrast and visibility of details. Phase contrast has a considerable advantage over darkfield in that internal specimen details are contrasted, whereas in darkfield the contrast is largely applicable to whole specimen versus the background and/or its surface details.

1. <u>The basic system</u>: Along with the optical path differences that are inherent within the specimen, further phase changes are

PHASE CONTRAST

introduced into the microscope optical system which, when combined with the specimen light phase changes, render the object visible. The contrast of a phase image is derived from the difference of the phases of the components of the light involved in the optical system and the specimen.

In the phase contrast microscope, the illumination is restricted to a hollow cone by an annular diaphram beneath the substage condenser. The condenser images the annular diaphram at infinity and the objective images it at its back focal plane. At that point a phase shifting element is placed. It is usually a glass disc with an annular ring of deposited metal which provides reduced intensity and a quarterwave phase shift to undiffracted light rays. The diffracted rays, emanating from the object itself, pass through the glass portion of the phase shifting ring. When the diffracted light and the phase shifted undiffracted light interfere at the image plane, the resultant is an intensity (amplitude) variation proportional to the phase differences of the interfering light waves. Refer to Figure 38 for a diagrammed illustration of the method.

2. Practical considerations

   a. Objectives: Phase objectives can be obtained in a number of corrections — achromats, apochromats etc. They have built-in phase plates located in the back focal plane which are not interchangeable. They are normally provided with matching substage annuli in sets.

   Although the phase objective does not have the resolving power of comparable objectives for conventional brightfield work, the improvement in contrast conditions provides a visibility condition that, at least for some types of specimens, is far superior in gathering total information microsopically.

   b. Centering: The phase annuli used in the substage must be matched with a particular objective equipped with a matching phase plate. For instance, a 10X phase objective and a 40X phase objective will require different matching annuli. The matching of the annulus to the phase plate allows exact centering to be accomplished and full advantage of the phase contrast to be exploited.

   If centration of the stage annulus is not exact in relation to the phase plate annulus, contrast will be much reduced. For that reason a "centering telescope" is ordinarily provided that is used in lieu of an ocular to view the back focal plane of the objective.

104 METHODS IN ACHIEVING AND IMPROVING CONTRAST

Both the annular diaphram and the phase plate are in view in the centering telescope.

Matching of the annulus to the particular objective in use is usually accommodated by a turret arrangement in the substage which allows the proper annular diaphragm to be selected. Single annulus phase condenser mounts are also available. This, of course, requires removing and inserting different phase annuli as the objective is changed. This type of mount provides for centration of the annulus also.

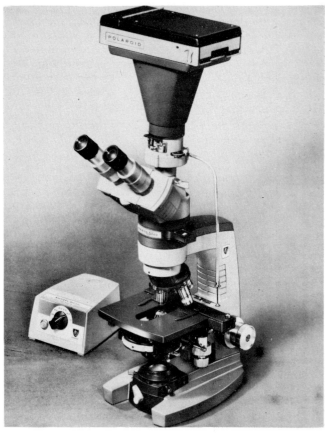

Figure 37. Phase contrast microscope with substage annulus turret. (photo courtesy American Optical Croporation.)

PHASE CONTRAST

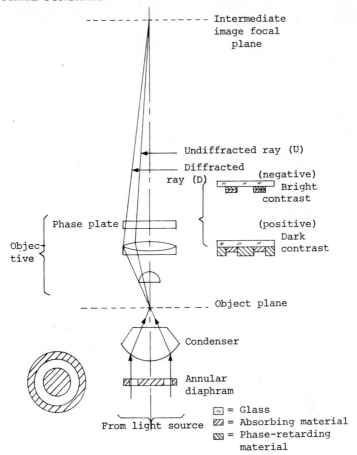

Figure 38. Phase contrast.

Adjustment is made first (with annular diaphram matched to the objective) by focusing on the specimen with the ocular in place. Then, with the ocular replaced by the centering telescope, the telescope is focused on the phase plate annulus. Then the substage annular diaphram annulus is brought into coincidence with the phase plate ring, usually by finger pressure. Once this is accomplished, the centering telescope is replaced with the ocular.

Some manufacturers provide a "viewing unit" that is used

instead of a telescope. This allows the normal ocular(s) (of a binocular, for instance) to be used when making centering adjustments. American Optical is one of the manufacturers providing such a feature.

      c <u>Illumination</u>: The method of illumination used in phase contrast is Köhler illumination. Adjustment of this type of illumination is simplified since the annular diaphram in the substage condenser can be observed by the use of the centering telescope usually supplied with phase contrast outfits. The image of the lamp field iris should be as large as the diameter of the annular diaphram.

    Although white light can be used the best results will not be obtained with it. Theoretically, monochromatic light of the same wavelength as used for the original calibration of the phase plates should be used if the full phase shift effect is to be realized.

    Since the phase plates are designed for phase shifts of one-quarter wavelength in the green portion of the spectrum, a green filter is necessary to take full advantage of the annulus. If white light is used, for instance, extinction by interference will not be complete for all colors and contrast will not, therefore, be as good. For critical observation and/or photomicrography the use of a green filter is mandatory.

      d. <u>Positive and negative contrast</u>: There are two manifestations of phase contrast obtainable, positive and negative, sometimes referred to respectively as dark contrast and bright contrast.

    In dark contrast (positive), the portions of the specimen having a larger phase difference are generally seen darker than those having small phase differences. As a result, a dark contrast objective, in most cases, produces an image corresponding to the ordinary brightfield and serves for examining minute details in the specimen.

    The mechanism whereby such an effect is accomplished is through control of the thickness of the film(s) contributing to the phase shift. If it is such that the directly transmitted light passing through the phase altering element is effectively accelerated over the diffracted light by one-quarter of a wavelength, the resulting contrast is positive or dark and the phase altering element is said to be accelerating.

    Negative or bright contrast is produced by retardation of the undiffracted light, regions of greater optical path in the specimen

PHASE CONTRAST

appearing bright against a dark background. The image is thus similar to a darkfield image.

In dark contrast (positive) the results obtained have a greater similarity to the usual hematoxylin-stained brightfield image and, perhaps for that reason, positive contrast images are more "popular" among microscopists. Also, this type of phase contrast provides greater enhancement to contrast and graded variations.

Since most microscope specimens exhibit irregular structure, the diffraction spectra are not clearly defined and the direct image of the illuminating source only is sharply defined. This direct light is, in addition, much more intense than the diffracted light. Since the direct rays always participate completely in the image formation, their intensity would tend to swamp out those from the weaker secondary diffracted rays. The relative light between the two (which includes both phase and intensity) determines the amount of contrast available. Compensation for this imbalanced distribution of light is accomplished by a light absorbing metallic film combined with the phase element to reduce the intensity of the direct rays. This absorption tends to equalize the intensities of the two interfering components and thereby increases the contrast sensitivity by allowing the weaker diffracted rays a greater proportion of influence on the final image formed.

By varying the transmission properties of the annulus, there can be produced a complete range from no contrast to pseudo-darkfield. The best contrast is somewhere in between, providing a gradation of darkness for the varying degrees of densities within the specimen. Brightfield gives no contrast at all unless the specimen is thick or absorbs strongly, and darkfield gives the extreme black and white contrast, occluding everything inbetween.

Some phase contrast objectives have been designed to provide a "good" contrast balance for either a dark or bright contrast image, as described previously. There are also available phase contrast objectives in sets which provide for a number of magnifications, and for dark and bright contrast conditions ranging from "dark-low-low" to "dark-medium" to "bright-medium". Such a wide choice of contrast variation can give better results when variously examining stained specimens, specimens containing large and small phase differences intermingled, fine fibers, granules, particles, protozoa, blood corpuscles etc.

F. ANOPTRAL PHASE CONTRAST: A very similar system of phase contrast to the former (Zernike) was developed by A.

Wilska. In this system, instead of a separate phase shifter placed in the back focal plane of the objective, he produced a heavily light-absorbing, nonreflecting area on one of the surfaces of an objective lens element. This serves as the phase shifting area for undiffracted rays. Instead of vacuum deposited metal, he originally used soot which was claimed to produce a "halo-free" phase contrast image. Reichert and other manufacturers currently produce instruments with anoptral contrast capabilities.

G. HOFFMAN MODULATION CONTRAST SYSTEM: This system was invented by Dr. Robert Hoffman and only recently has been demonstrated publicly. It operates without phase contrast objectives or interference contrast accessories. The following performance features are claimed for the system.

(1) Reveals unstained transparent or living specimens which are invisible in ordinary brightfield microscopy. The modulation contrast system creates light and dark contrast of even the most minute rounded portions of invisible specimens. The nonflat regions are referred to as phase gradients.

The system converts phase gradients into intensity variations, one gradient is dark, the opposite gradient light. All non-gradient regions (flat or of uniform refractive index) appear as gray. Light intensities in the image are modulated above and below an average gray value, thus the name modulation contrast.

(2) Creates 3-dimensional images of both transparent and opaque objects.

(3) Provides optical sectioning of any specimen with no obscuring effect from structures above and below the plane in focus.

(4) Does not require elaborate equipment changes in standard microscope optics. Two basic parts comprise the system, a modulator and a slit aperture. Optional variable contrast as an accessory is available.

(5) Eliminates halo to permit viewing of minute details.

Figure 39 illustrates the principle of operation.

H. INTERFERENCE CONTRAST: Contrast in interference microscopy is produced by the effect on the interference phenomena of the path differences caused by the optical thickness (for reflecting objects) of the specimen. This method of contrast enhancement makes use of the same properties of the specimen as

INTERFERENCE CONTRAST

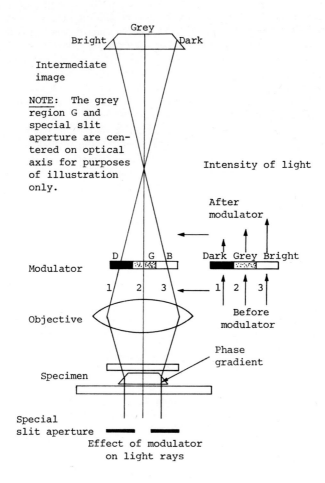

Figure 39. Modulation contrast. (drawing adapted courtesy Modulation Optics Inc.)

does phase-contrast microscopy. Advantages of the interference method are that "phase-contrast halo" is avoided, measurement of path differences can be precisely made and the nature of the contrast can be continuously varied by altering the overall path difference between the two interfering fields to obtain the best visibility condition. Also, in some types of interference microscopes, the numerical aperture of the objective need not be restricted to a maximum of 1.0.

There are many ways in which the interfering fields are obtained. Generally, one field contains the image of the specimen and is termed the "image" field, the other field differs from the image field in some way (and is called the reference field), but the two fields are ultimately mutually coherent at every point. Complete coverage of interference microscopy is the subject of another book in this series. A brief treatment of two of the more common schemes in use follows.

1. <u>Jamin-Lebedeff system</u>: This particular system of interference microscopy will be gone into only in sufficient detail to delineate its action to provide contrast and to differentiate it from that of the differential interference contrast system to follow.

Essentially this system is the combination of a polarizing microscope and a two-beam interferometer, wherein one beam traverses the specimen and the other does not. The essential components necessary for accomplishment of interference effects with this system are:

(a) A light source with condenser — adjustable for Köhler illumination.
(b) Polarizer (mounted in the filter holder of the condenser).
(c) Substage condenser aperture diaphram.
(d) Substage condenser.
(e) Beam splitter.
(f) Beam combiner.
(g) Microscope objective.
(h) Analyzer.
(i) Eyepiece or ocular.

This particular method of interference microscopy is used primarily in making measurements of phase shifts (phase changes or optical path variations of down to less than 1/200 wavelength) in determinations of refractive index (R.I.), thickness, concentrations, wet and dry mass of cells, cytological components such as protein and water, bacterial size etc. In doing so, excellent contrasts are provided in the image. Variable intensity contrast is readily obtainable when using monochromatic (or properly filtered) light and variable color contrasts are produced with white light.

  a. Zeiss interference microscope: This system is essentially Jamin-Lebedeff derived and operates briefly as follows (refer to Figure 40).

A polarizer produces plane polarized light such that it is incident upon a birefringent crystal plate (beam splitter) cut obliquely to its crystallographic axis. Light waves passing through split into two beams (parallel to one another) through the object space at a definite distance from one another. One is called the "measuring beam", the other the "comparison beam". In addition, a half-wave plate is provided with the beam splitter which rotates the vibration plane of the plane polarized light leaving the beam splitter, through 90°. (This holds true for only a specified wavelength of light — in this case 546 nanometers.)

The beam splitter and the half-wave plate are located in front of the front lens of the substage condenser. The beam combiner is a birefringent crystal plate identical with the beam splitter in thickness and optical orientation. It combines the measuring and comparison beams and is mounted in front of the objective front lens. It can be rotated azimuthly together with the objective from the exact diagonal position of its principal directions of vibration to the crossed polarizer and analyzer. This forms an adjustment of the interference system to provide a variation of high contrast (bright, dark or colored).

The illuminating and imaging beam paths of this interference microscope, for transmitted light in principle, are the same as those for a polarizing microscope.

The optical elements for bringing the light into a suitable state of polarization for interference are placed between the luminous field diaphram and the tube rest. These elements are polarizer, beam splitter, beam combiner and analyzer. For quantitative determinations, a compensator can be introduced below the analyzer or above the polarizer.

  b. A-O Baker interference microscope: Another example of a system based on ideas of Jamin (1868) and Lebedeff (1930) is the A-O Baker Interference Microscope. Reference is made to Figure 41.

It will be noted that the condenser has a double refracting

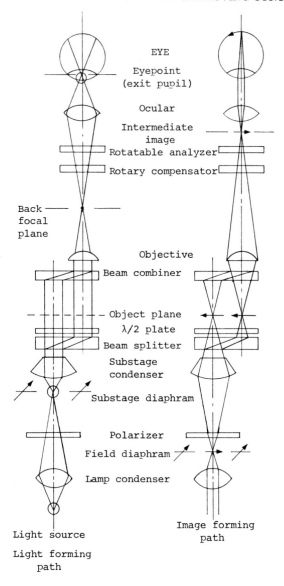

Figure 40. Interference microscope (Zeiss). (adapted from Zeiss illustration.)

INTERFERENCE CONTRAST

plate which divides the light into two beams. The objective has a corresponding plate which recombines the beams after one of them has passed through the specimen. Below the condenser is a polarizer and above the objective is a quarter-wave plate and an analyzer.

The polarizer presents plane polarized light at 45° to the axis of the double refracting plate of the condenser. This plate separates the polarized light into two beams which are plane polarized at right angles to one another. One beam passes through the specimen and undergoes a relative phase shift dependent upon the optical-path difference between the specimen and its surround. The other beam (reference) focuses above or to one side of the specimen focus. The double refracting plate in the objective recombines the two beams and transmits them to the quarter-wave plate above the objective which changes the two oppositely polarized beams into counter-rotating, circularly polarized light. The resultant of the two circularly polarized beams is plane polarized light whose orientation depends on the phase difference between the circularly polarized beams. Phase differences can be observed through the ocular in selective color contrast with white light illumination, or in variable dark-to-bright contrast with monochromatic or properly

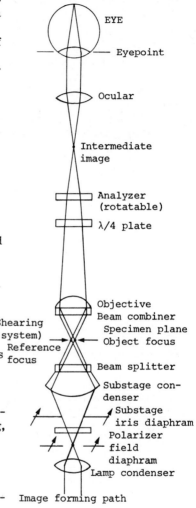

Figure 41. A-O Baker Interference microscope. (adapted from illustrations courtesy American Optical Corporation.)

filtered light by rotation of the analyzer.

The A-O Baker instrument may be used in either of two modes of operation.

(1) <u>Shearing system</u>: In the shearing system, one beam of the divided polarized light focuses through the specimen and one passes to one side of the specimen (refer to Figure 41). As a result, the specimen and reference fields are separated laterally and an out-of-focus, or sheared, image of the specimen will be seen to one side of the specimen. This allows for critical measurements of separated features, such as single cells.

Complete lateral separation between the true image and the sheared one can be achieved provided the true image is smaller than the lateral displacement of the sheared image for the field of the objective in use. With the 10X objective system the center of the sheared image, or reference field, is typically separated from the optic axis by approximately 330 micrometers; 160 micrometers for the 40X system and 27 micrometers for the 100X immersion (water) system. Therefore, objects of up to 330, 160 and 27 micrometers, respectively, in diameter can be viewed without any overlap between the object and its reference field.

(2) <u>The double-focus system</u>: In the double-focus system, the reference field is an out-of-focus image of the object plane (refer to Figure 42), the reference being displaced longitudinally on the optic axis of the microscope. This makes the double focus system particularly well adapted to the qualitative examination of extended specimens such as tissue sections. Also, the reference beam is less affected by an inhomogeneous surround than the shearing system, an advantage in variable contrast microscopy. Even though this latter statement is true insofar as it goes, the instrument is still primarily, even in this mode, a measurement device as with 10X and 40X objective systems the diameter of the reference area surrounding the feature is 90 micrometers, and with water immersion (100X) there is a circular reference area of 20 micrometers diameter.

2. <u>Differential interference contrast (DIC)</u>: To obtain a superior variable contrast image, and one in which the inhomogeneity of the surround plays a minimal part, differential interference contrast (DIC) is used. The major difference between DIC systems and the previously described Jamin-Lebedeff systems is in the physical spacing or separation of the two beams of light traversing the specimen area (refer to Table VII). Where the

INTERFERENCE CONTRAST

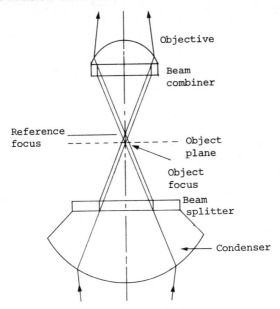

Figure 42. A-O Baker double-focus system. (adapted from illustration courtesy American Optical Corporation.)

TABLE VII

A comparison of interference contrast systems

Jamin-Lebedeff			DIC		
Obj.	Condenser	Shear in object plane	Obj.	Condenser	Shear in object plane
10X NA 0.22	10X match	560 μm	16X NA 0.32	setting I	1.33 μm
40X NA 0.65	40X match	180 μm	40X NA 0.65	setting II	0.55 μm
100X NA 1.0	100X match	56 μm	100X NA 1.25	setting III	0.22 μm

Jamin-Lebedeff based systems are usually limited to a numerical aperture of 1.0 (because of the large shear), the DIC systems can utilize the full numerical aperture of oil immersion objectives (1.25 etc.).

It will be noted (referring to Table VII) that the lateral displacement of two sheared wavefronts in the plane of the object is less than the resolving power of the microscope objective. Therefore, no double contours will be visible in the image whatsoever.

Referring to Figure 43, we see two sheared wavefronts in the image plane, illustrating a bossed specimen (one with a raised or elevated feature). According to the illustration, it is assumed

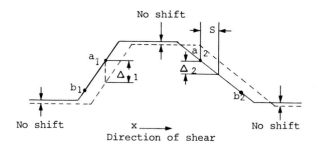

Figure 43. Differential interference contrast (two sheared wavefronts).

that the path difference between the sheared images is $\bar{\Delta}$. The phase shift then amounts to the products of the shearing distance in the image plane multiplied by the differential coefficient of $\bar{\Delta}$, $\frac{\partial \bar{\Delta}}{\partial x}$, and therefore over the part $a_1 b_1$

$$\Delta_1 = S \cdot \left(\frac{\partial \Delta}{\partial x}\right)_1, \quad (S = \text{shearing distance})$$

and over the part $a_2 b_2$

$$\Delta_2 = S \cdot \left(\frac{\partial \Delta}{\partial x}\right)_2$$

where x denotes the direction of shearing. The path difference of the two rays, for instance, is equal to the product of the shear and the differential coefficient of the wavefront, hence the name differential interference.

Therefore, it is obvious that the brightness variation in monochromatic light, or the interference colors in white light indicate a phase difference proportional to Δ_1 at part $a_1 b_1$ and a phase difference proportional to Δ_2 over $a_2 b_2$.

In flat portions of the specimen (perpendicular to the optical axis — indicated by "no shift") the phase difference will be zero since the slope of the surface is zero, and will be manifested in the image as black in either monochromatic or white light. However, in white light any parts with a phase slope ($a_1 b_1$ or $a_2 b_2$, for instance) will appear colored against the black background. The color is determined by the phase difference in each case.

Referring to Figure 44, the essential parts of such a system are very similar to the Jamin-Lebedeff systems. A light source, condenser, beam splitter, polarizer, objective, beam combiner, analyzer and ocular are used.

As mentioned previously, the major difference between the two systems is the difference in beam separation. The Jamin-Lebedeff systems use beam splitters and combiners (double-refracting plates) that separate the beams sufficiently such that accurate measurements may be made on comparatively large microscopic features or specimens. The design is essentially directed, by this means, at measuring capability, with the contrast feature a secondary one. The measurement feature is accomplished by adjustment of the orientation of the beam splitter — combiners with respect to the optical axis.

In the DIC systems, the beam splitters and combiners are designed such that they only separate the two beams (and recombine them) by distances which are less than the resolving power of the microscope optics. Here, measurement is a secondary feature and design is directed toward variable contrast capability at maximum resolution.

The two systems are, therefore, complementary to one another and not competitive or contradictory at all. Adjustment of the position of the beam splitters and combiners (with respect to the optical axis) accomplishes the variable contrast feature, with the measurement capability limited to interpretations of color values compared with standard color scales.

For instance, if the second beam splitter (or combiner) is

tilted with respect to the optical axis (out of perpendicularity), the path difference between the two wavefronts (Figure 44) is increased or decreased, and different interference colors will be produced.

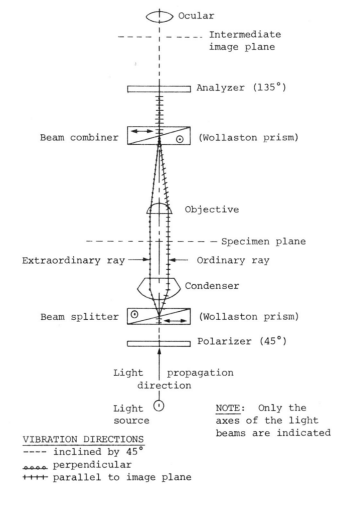

Figure 44. Nomarski differental interference contrast.

INTERFERENCE CONTRAST

This kind of adjustment, then, is used for effecting different color contrasts. If the background is made dark, an interference image similar to the bright contrast in phase contrast microscopes is obtained.

When the background is made gray, slightly lessened from the darkest, the image will show a gray "hypersensitive" color, giving the highest contrast. This appears in a relief picture of the phase difference distributed over the entire specimen.

When the background is made a sensitive purple, the interference colors of the specimen will depend upon the inclination of the optical thickness when the specimen has a difference in level.

When the background is made sky-blue, there appears an interference image similar to the image produced in the phase contrast microscope which has a phase plate that is least absorptive.

The adjustments of the particular beam splitter (or combiner) to accomplish the above effects are sometimes termed "compensation" adjustments. They give rise to high contrasts in the case of specimens of small phase differences. For specimens of great phase differences (with considerable elevations and depressions) a background of other interference colors (provided by tilting the phase splitter to the optic axis) will possibly provide better color contrast.

a. Nomarski differential interference contrast: This is a specific practical system in common use which is available in both transmitted and reflected light versions. In this method incoherent illumination at high NA is usable which results in exceptional image brilliance and with resolution almost twice as great as in phase contrast.

Referring to Figure 44 it will be seen that essentials of all interference type instruments are used. The beam splitter and beam combiner, typical of this system, are Wollaston prisms. Since both beams are separated by a distance less than the resolving power of the optics, the terms "specimen beam", "reference beam" etc. have no meaning in this system, as is the case with other DIC systems.

Because of the use of Wollaston prisms instead of the plane-parallel plates of the Jamin-Lebedeff based systems, the beam splitting is angular instead of lateral. The ordinary ray departs from the interface of two pieces of uniaxial material, such as quartz or calcite, at an angle divergent from the extraordinary ray. The two pieces of uniaxial material are cemented together such that their axes are at right angles to one another. When a plane polarized wave (bundle) of light, whose vibration plane is 45°, is perpendicularly incident upon the Wollaston prism, it will be split into two plane polarized waves in the lower prism whose vibrating planes will each form an angle of 45° with the incident wave.

The Wollaston prism is particularly useful where intensities in polarized light are involved since the images of the two beams, whose vibrations are perpendicular to each other, can be compared side by side.

In the Wollaston, light enters normal to the surface and travels perpendicularly to the optic axis until it strikes the second prism where double refraction takes place. It deviates both rays and consequently yields a greater separation of the two chromatic beams, an advantage in a microscope system largely devoted to interference contrast in color.

The interference colors can be varied to suit any object detail from dark to bright contrast. There is an illusion of increased depth of field (focus) due to the high NA which limits the depth in which the interference contrast is produced. Because the contrast is produced in such thin optical section, the interference image is free of out-of-focus details and produces an illusion of greater depth of field. A further enhancement of the appearance of depth of field is created by a strong shadow-cast effect, especially with the adjustment for the gray "hypersensitive" color mentioned previously. A three-dimensional effect is a first impression when this type of interference contrast is initially encountered.

In phase microscopy, phase details are made visible due to differences in refractive index or thickness in the specimen. When there are uniform phase details only, areas containing steep gradients of refractive index will appear as different intensities (contrasts) in the final image.

Phase details by Nomarski appear as apparent relief or shadow-cast images. The background in the Nomarski system is in the area of the specimen itself or, in the case of very small objects, the microslide where no object is present. With white light illumination, the background can be made to appear colored, black and white or gray. Therefore, color effects can be produced whether there is an object in the light path or not. This, of course, is a powerful tool for optical staining in many variations. As optimum contrast is achieved in the case of structures running diagonally to the field of view, a rotatable microscope stage is of considerable advantage.

INTERFERENCE CONTRAST

In reflected light applications, Nomarski DIC is very useful in surface examination of an object that cannot be, or should not be, stained or etched. Nomarski renders a sharply defined relieflike image of an opaque specimen with excellent contrast in a wide range from zero order to high sensitive tints. Figure 45 shows the schematic of a Nomarski-Epi system by Nikon. Note that only one Wollaston prism is necessary, as it is traversed twice through the action of the half-mirror. One time the Wollaston prism acts as the beam splitter and the next, on the return of the light information from the specimen through the objective, as a beam combiner.

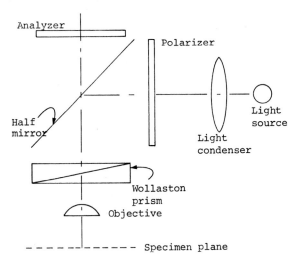

Figure 45. Nomarski differential interference contrast (reflected light).

I. SCHLIEREN MICROSCOPY: Schlieren microscopy is based on a refraction technique that makes visible small index of refraction gradients in a microscopic sample.

A basic schlieren system includes a point light source, a strain free high quality, long focal length lens, an obscuring spot or knife-edge, placed precisely at the light source image position, and a viewing telescope focused on the sample which is placed near the schlieren lens. Since the light that enters the telescope must pass the obscuring spot, the field will be dark if the imaging is perfect and no sample is present.

A practical compact microscope based on the principles above is shown diagrammatically in Figure 46. It will be noted that a transfer lens is employed below the microscope stage which images the preparation to be examined at a point such that it can be further magnified by the microscope optics.

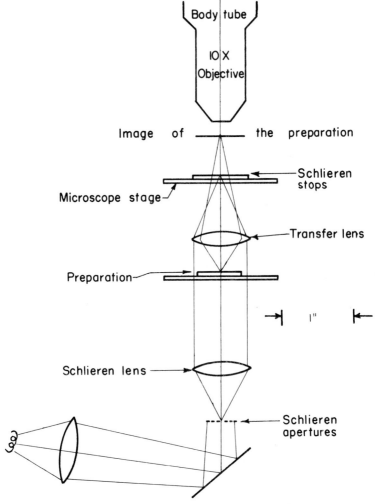

Figure 46. Schematic schlieren microscope. (drawing courtesy Walter C. McCrone Associates, Inc.)

Schlieren effects result from deviation (refraction of light) around the schlieren stop by refractive index gradients or thickness variations in the specimen. The effect is only apparent when the optical path differences are very small and usually provide black and white images unless the index of refraction difference is comparatively great when the image will show color tints.

The schlieren microscope is, in effect, a very sensitive dispersion staining device. This can be attributed to the small size of the schlieren apertures (10-12 μm in diameter) compared with the single dispersion staining aperture (2-3 mm in diameter). Much less deviation of the light beam is necessary to miss the 10-12 μm schlieren stops than to miss the 2-3 mm dispersion staining stop. As a result dispersion staining colors can be readily observed with the schlieren scope for systems, e.g., KCl in Cargille liquid 1.490, that show no colors with the usual dispersion staining objective. The schlieren microscope also functions as an interference microscope giving interference colors related to optical path differences for adjacent detail in the specimen. This is apparently due to the fact that diffraction occurs at the edges of each of the hundreds of 10-12 μm schlieren apertures and the diffracted beam travels a different path through the preparation than the direct beam. On recombination of the diffracted and direct beam in the image plane interference then results.

The use of such an instrument has only thus far been barely touched upon. Preliminary investigations show that it will prove to be very useful in many fields. Studies of liquid diffusion processes and crystal growth and decay are areas that can definitely benefit by its use.

A practical instrument of this type is available from Walter C. McCrone Associates, Inc. The necessary accessory parts are built in and/or attached to a basic biological microscope. The usual microslide preparation is placed on a new small stage located in the substage of the basic instrument. The schlieren stops are held in a mechanical stage on the original stage. The bodytube is focused on the image of the preparation or the schlieren stops. Figure 47 photographically illustrates the essential spatial arrangement.

J. REFERENCES AND COMMENTARY: A number of the "methods" herein described for obtaining contrast or for improving contrast under certain conditions are, or are becoming, microscope systems of major importance. Many of them are treated extensively in the literature and a number are the subject of

Figure 47. Schlieren microscope. (photo courtesy McCrone Associates, Inc.)

REFERENCES AND COMMENTARY

entire books in this series. The references following are representative only and selected for their more direct applicability to material in this chapter.

1. Jenkins and White, Fundamentals of Optics, 2nd edition, McGraw-Hill Book Company, 1950.

Information on various prisms, including the Wollaston prism used in Nomarski DIC.

2. Meyer-Arendt, J. R., "Schlieren method for mass determinations in microscopic dimensions," Rev. Scient. Instru. $\underline{28}$ (1), 28-29 (1957).

3. Dodd, Jack G., "A schlieren microscope," The Microscope, 15, 5 (July-August 1966).

4. Dodd, Jack G., "Observations with a schlieren microscope," The Microscope, 17, 1st Quarter (January 1969).

5. McCrone, W. C., "A compact schlieren microscope," The Microscope, 20, 4th Quarter (October 1972).

6. Taylor, R. B., "A relief contrast system," The Microscope, 15, 4th Quarter (October 1967).

A system similar to Anoptral contrast — depending upon an annulus and a ring of light absorbing material in the objective is described.

7. Meijer, D. J. S., "Wire or shadow contrast," The Microscope, 11, 2 (September-October 1956).

A wire or hair is placed in the microscope tube (at right angles to the optical axis) and a disc with a slit is placed in the substage filter carrier.

8. Clark, (Ed.), The Encyclopedia of Microscopy, Rheinhold Publishing Corporation, 1961.

Appropriate sections treating topics included in this chapter.

9. Jackson, Allan, Amateur Photomicrography, 7th edition, The Focal Press, 1958.

Information on liquid filters and instructions for their composition, both for visual and photomicrographic use.

10. Photomicrography — A Handbook on Photography With the Microscope, Eastman Kodak Company, Fourteenth Edition, 1944.

This version is no longer in print, but contains very good treatments of many of the topics included in this chapter. Especially good on darkfield illumination and on color filters. Although primarily directed to the photomicrographer, the basic information in this little book is excellent background and guidance for any light microscopist.

11. Hoffman, R. and L. J. Gross, "Reflected-light differential interference microscopy; principles, use and image interpretation," Microscopy, 91, 149-72 (1970).

12. Pluta, M., "A new polarization interference microscope," The Microscope, 18, 113-22 (1970).

13. Smith, R. F., "Color contrast methods in microscopy and photomicrography," — Part I, Part II, Part III, Phot. Appl. Sci. Technol. Med., 4, 16, 24 (1970); 5, 19-24, 36 (1970); 6, 19-23 (1971).

14. Mihajlovic, A., "Some comments about interference contrast in quantitative and color microscopy," Metallography, 5, 1 (1972).

15. Lang, W., Nomarski Differential Interference Contrast Microscopy III, comparison with Phase Contrast Method, Zeiss Information 18th year, 75/76, 22, 1971.

16. Ono, A., H. Hashimoto and A. Kumeo, "Application of darkfield technique to biology," Jeal. News, 10e, 2 (1972).

17. Hoffman, Robert and Leo Gross, "Modulation contrast microscope," Applied Optics, 14 (May 1975).

18. Hoffman, Robert and Leo Gross, "The modulation contrast microscope," Nature, 254, No. 5501, 586-588 (April 1975).

CHAPTER 3

SPECIMEN PREPARATION AND OBSERVATION

A. INTRODUCTION: There are a considerable number of excellent and authoritative references devoted to specimen preparation available to the microscopist. They offer complete laboratory procedures, detailing the necessary steps to bring the object to a condition suitable for microscopical examination. Also, there are numerous detailed sources describing microscopical methods and instrumentation.

This chapter will include a discussion on the purposes of the various steps of ordinary preparation, the factors resulting therefrom and their possible effect on image interpretation. In addition, special methods of preparation and observation will be briefly described to provide a more complete overview of this important aspect of light microscopy. It is hoped that the contents of this chapter will provide a better understanding of the specimen preparation process, and furnish information both complementary and supplementary to that of other more specific and detailed treatments of the subject.

B. KILLING AND FIXING: The first step towards bringing the specimen material under the microscope for examination often entails killing and/or fixing.

Living material, animal or plant, is sometimes observed in its natural state, but extended examination usually requires that it be killed and further processed. The purpose is to render the material suitable for detailed and analytical observation with the microscope and/or to provide a permanent record in the form of microslide preparations, or to prepare it for photomicrographic recording. During these preliminary steps the object is to preserve the tissues or cells in as lifelike a condition as possible, and to make that condition durable enough such that it remains unchanged throughout any other treatment such as staining and mounting. Often the two procedures (killing and fixing) are simultaneous. Sometimes narcotization of living animal material is necessary so that it is in a relaxed condition before killing. Fixing is often accomplished by treatment of the killed material in alcohols.

An important factor in fixing is that in order to preserve cell shapes etc. the osmotic pressure between cell contents and external fluids needs to be equalized, thus keeping the cell (or other structure) from expanding or contracting. Two fixing agents having opposite effects can be used in a properly proportioned mixture to accomplish this. Chromic and acetic acids are used, for instance, in botanical work. Further fixing or preserving is quite often then accomplished by alcohols, as mentioned above, or by a preservative such as a weak formalin solution.

1. <u>Artifacts</u>: In specimen preparation, every step should be considered in light of what artifacts are produced, such that proper interpretation of the microscopical image is made; otherwise the true nature of what is seen may be obscured or distorted and incorrect conclusions drawn thereby. In the killing and fixing of living specimens the ideal situation is to render the object in a lifeless state exactly as it was in life, yet suitable to be further treated and examined without change. That this is a very difficult matter, to say the least, is recognized. It is seldom accomplished with perfection.

If the organism is insufficiently "relaxed" and in an unnatural agitated or tense state when killed, that will often be reflected in the final preparation. This will appear in whole mounts as contractions and distortions in the position of the main body and limbs. Also, the cells of the body, plant or animal, may be under extreme pressures or abnormal chemical conditions in such agitated or abnormal states, and their true natural condition thus not revealed in the final preparation.

In the fixing phase, even if the killing is successfully performed, the individual cells, their configuration, size and contents, may become distorted, such as to be unnatural and difficult of proper interpretation. Some shrinkage and/or swelling of cell walls may take place in even the most careful fixing procedures. Improper dehydration, either underdone wherein remaining water will adversely effect other preparation steps, or hardening to the extent of brittleness and distortion, must be recognized as possible artifacts of this stage of specimen preparation.

C. <u>SECTIONING</u>: Because much material for examination by the light microscope is usually far too thick and opaque, the next step is to section it. This procedure provides "slices" of material thin enough to be examined by transmitted light. Plant sections are usually about 40 micrometers thick, and animal organs 20 micrometers or thinner. Inorganic materials such as rocks and minerals etc. are also examined in section.

1. <u>Cutting</u>

a. <u>Animal and plant tissues</u>: Sectioning is accomplished either by "freehand", using a razor or by the use of a microtome.

SECTIONING

The microtome may be a simplified hand microtome, or a rotary or sliding sledge type. The more sophisticated microtomes can provide sections as thin as 1 micrometer.

Before the specimen material can be sectioned it must be supported in some manner such that during the sectioning process it is not distorted by forces exerted by the cutting edge of the knife. This usually means supporting the specimen material both internally and externally. The term "embedding" is quite often used to describe this preliminary step to sectioning, and is understood to include the infiltrating of tissues and structures for internal support as well. In only exceptional cases can objects be cut by a microtome without some form of embedding treatment.

Two treatment methods employed are saturation of the object with paraffin or with celloidin. The paraffin method is the quickest and most common. As paraffin is soluble in water, alcohol dehydration of the object is a prerequisite to embedding by this method. Freezing before sectioning the embedded material also is sometimes used with success (although sections cut from frozen embedments do not show fine details as well).

b. <u>Other materials</u>: Materials other than the usual plant or animal tissues are prepared in different ways dependent upon their characteristics. Some of them which are too hard or otherwise unsatisfactory for cutting initially with a knife can be pretreated such that they can be cut.

c. <u>Wood</u>: Very thin slices or sections of wood, longitudinal, transverse or radial, are commonly required for microscopical examination. When fresh, most woods can be easily cut with a blade of some sort, otherwise special preparation methods are used. Soaking such material in hot water or in an alcohol-glycerine mixture (equal volumes of each) for several days is a common procedure and renders easier cutting of very thin sections. The cutting may be done by hand with a razor or a very sharp wood plane.

d. <u>Fibers</u>: Because of their flexibility in general, fibers require some special support during sectioning, either transverse or longitudinal. A common procedure is to draw bundles of fibers through a very small hole in a thin metal plate and to make transverse sections by drawing a cutting edge coincident with the metal surface across the bundle. The hole usually is about 0.5 mm in diameter and cutting is done on both sides of the plate with a razor. Specialized microtomes are sometimes used in crossectioning fibers. One of the most common is the Hardy microtome, consisting of two rectangular metal plates held together by lateral ridges. Shearing action of the plates accomplishes the necessary section cuts. Embedding of fibers in celloidon or gelatine, for instance, then freezing and cutting in a conventional microtome is also sometimes done. The latter method is often used in making longitudinal sections of fibers.

e. Bone: Cutting materials which contain lime such as bones or animals with calcareous shells etc. is made possible by decalcification. The material is first fixed and then treated by special acid mixtures. A good decalcifying agent is trichloracetic acid which decalcifies rapidly and hardly changes the tissue. Time required for treatment varies with different materials until it becomes sectile enough to be cut readily with a knife.

2. Grinding

a. Rocks and minerals: Making thin sections of very hard materials is usually done by grinding. Small chips of rocks, minerals or other such materials are cemented to microslides and the upper surface ground smooth on a flat metal or glass plate with increasingly finer grades of carborundum or similar grinding compounds. When the upper surface is flat, the chip is removed by solvent action on the cement and remounted, flat side down. Then final grinding in a similar manner, but to a much finer degree, is continued until the material is of the proper thickness. Frequent examination of such sections under the microscope, during the grinding process, controls the quality and thickness of the section.

Similar grinding procedures are also used to expose chambers in microscopic shells or tests of microfossils. The grinding is continued until the section or specific attributes of the object are revealed.

3. Corroding: Sectioning or exposing certain chambers or details hidden by overlying material can sometimes be more selectively accomplished by corroding.

Materials that are particularly susceptible to this treatment are calcareous microfossils such as foraminifera. For example, an individual test is mounted in paraffin on a glass slide, being completely submerged. Then, under a stereomicroscope, a particular area to be corroded is carefully cleared by removing paraffin with a fine needle. With a very dilute acid solution applied with a fine brush or bristle, even individual chambers in the test may be gradually opened up to view.

4. <u>Artifacts</u>: In cutting and/or grinding operations on specimen material prepared for microscopical examination, artifacts of one kind or another are usually introduced.

When the knife used for cutting is insufficiently sharp, chattering or vibration of the blade can create serrations or "streaks" in sections. If the blade is damaged, a similar type of artifact may be introduced by an inhomogeneous cutting edge.

Insufficient support of the specimen will show up in extreme deformation of structure such as cell walls. This latter artifact is one of the most common in making thin sections. Recognition is easily made by comparison of the specimen in both embedded and unembedded conditions.

Cutting operations, dependent upon the hardness of the material and the speed with which they are performed, can introduce some sort of flow artifact. This is often exhibited by deformation at the cut surface which is not attributable to insufficient support of the material.

During grinding of thin sections of hard materials, "pulling-out" of portions of the material may occur, falsely indicating voids in the final section. This is usually circumvented by using very fine grinding compounds and proceeding slowly instead of rapidly. The nature of the supporting material (cement) is important in controlling this kind of artifact. Mounting soft materials in epoxy resins prior to grinding avoids this condition.

D. <u>STAINING</u>: After sections thin enough for examination by transmitted light are made, an improvement in contrast of various parts of a specimen may be accomplished by the use of stains. Objects or portions of objects impregnated by stains are rendered more visible in color, and with other portions stained in contrasting colors, much improvement in visibility is provided.

In the process of staining, the dye or coloring agent, becomes attached to the structure to be stained and remains attached throughout the various procedures involved in making the complete preparation and final mount. The most common reason for permanent attachment of the dye differentially is not mechanical adhesion, but is due to the fact that the charge on the dye becomes balanced by the charge on the particle or structure dyed. The degree of adhesion of the stain to the structure can be increased by the action of mordants. Sulfates are the most widely used mordants in microtechnique — particularly the double sulfates known as alums.

Other methods of differential staining are dependent upon the solubility of the dyestuff (oil soluble, water soluble etc.), or by simple difference in permeability between different types of tissue. Different sized molecules of colored dyes will permeate different tissues and/or portions of objects with different facility.

Dyes attached to tissues by mordants may be removed by ionized solutions — usually acid — or with the use of solutions of the mordant itself. This is, in addition, another basis for differential staining. It is also evident why acid-containing mountants are detrimental to stained materials. Many materials are susceptible to staining besides ordinary plant and animal tissues, including fats, skeletons and bones.

Staining may be either direct or indirect. In the direct method the coloring agent is applied to the tissue or object which is removed when sufficient dye has been absorbed.

In the indirect method the specimen material is soaked in the stain solution, then differentiated. Direct staining is considered preferable in the case of whole mounts, while the indirect method is employed on sections of small objects.

With the many different methods of staining, and the large number of stains and wide variety of objects susceptible to the technique, there has grown an enormous technology, and voluminous references are available on all aspects of it. Continuing research constantly expands the possibilities of staining products and application.

Stained objects are usually mounted in resins which are not soluble in water. A great many stains fade rapidly in the commonly used water-soluble mounting media. As natural resins (such as Canada balsam) are quite often acidic, stained material is usually best mounted in a synthetic medium with a refractive index equal to the refractive index of the mounted tissue, and which is neutral. This situation provides the best contrast for the colored dyes and assures their permanence.

Stained preparations are quite often illuminated differently than unstained ones. Since, in stained material, the contrast between structures is enhanced by a color difference, the substage condenser diaphram can remain open to a greater degree, thus providing the light necessary to the objective for greater resolution of fine detail. Unstained material, which is commonly mounted in a medium of a low refractive index (water or glycerine mixtures), sometimes requires that the substage condenser diaphram be closed down further for sufficient contrast (none being

SURFACE EXAMINATION

furnished by color differences) and thus possibly sacrificing some degree of resolution. Thus, stained preparations not only provide greater contrast by color, they also provide a better condition for maximum resolution by the optics involved. For the observation of stained specimens, in biological preparations, for instance, the ability to faithfully reproduce the variations in light absorption and thus take maximum advantage of the differentiation by color, it is very desirable that the objectives and condensers used on the microscope stand be of a high order of correction.

In the interpretation of stained materials, because of the vagaries of the human eye mentioned in the previous chapter, it may be necessary at times to apply more than one staining method to resolve such problems, which are especially troublesome at higher magnifications. In the practical sense, both size differences and color differences of details in a three-dimensional object may be met with and require some very critical analysis on the part of the microscopist to avoid erroneous conclusions.

Because the retinal cones of the human eye are insensitive at low illumination levels and the fact that their optimum sensitivity in a light-adapted eye is at about 588 nanometers, some considerations relative to those conditions need to be taken care of in staining certain types of specimen material. The quality of color impression changes with light intensity. Low intensities cause an eye sensitivity shift toward short wavelengths. For instance, yellow and red appear darker, and green and blue brighter. Therefore, objects which will necessarily have to be studied at high magnifications and low levels of illumination, such as bacteria, mitochrondria etc., are better stained with red stains than with blue or green because of the increase in visibility.

E. SURFACE EXAMINATION: The light microscope is used in the examination of surfaces of all kinds of materials: organic, inorganic, transparent or opaque. Even though an object may be examined by transmitted light in the more conventional manner, it often can be profitably examined in its surface features to reveal additional information. Opaque objects or materials and objects too large to be placed under the microscope are frequently subjected to surface examination by various means to great advantage. Some of the special means of preparing for this type of examination follow in this section.

1. Grinding and polishing: Probably the most commonly known method of preparation for examining surfaces is that of grinding and polishing of metal for metallographic work. The

same or similar methods are used for many opaque materials, including rocks, minerals and ceramics.

The material is usually mounted and/or imbedded in a cement to hold or support it during grinding and polishing operations. Bakelite was, for years, a conventional method for supporting opaque minerals. In recent times, however, various improved epoxy resins and plastics are in more general use.

Whatever method of support is used, the material is ground smooth with ever finer abrasives grading into the finest of polishing compounds. In grinding and polishing minerals, for instance, the specimen is first ground with 120 grade carborundum on an iron lap and finished off on a copper lap with 3F carborundum. Then it is mounted (in Bakelite or other supporting material) and fine-ground with 600 grade abrasives in several steps. Final polishing is with rouges or other polishing compounds on a cloth lap.

The polished surface of the specimen material is normally examined in vertical illumination. Identification of the opaque minerals by such preparations has become a routine matter. Properties of reflectance, polarization, color, hardness, paramagnetism, crystallization and crystal orientation are revealed by expert examination.

2. Etching: Polished surfaces, under vertical illumination, on examination by an expert in a specific field using such techniques, reveal considerable information, as mentioned above. Most metals and some minerals are sufficiently reflective to produce images of high contrast when etched — a common metallographic procedure. When this type of technique is an advantage, it becomes a succeeding step to grinding and polishing.

Although strictly speaking, the etching of polished surfaces of metals or minerals might be properly classified with "microchemical" tests, they are usually referred to as "etch" tests, being performed on the polished surface itself as distinct from "microchemical" tests performed on a glass slide.

The etching effects, or lack of effects, contribute information, microscopically, to the investigator which further enhances his knowledge of the material, including its consistency, structure and properties necessary for identification (minerals) or application in industry (metals). Dependent upon the material, the reagents used may be acid or basic. Their strength, method of application, duration of reaction and interpretation of results are far beyond the scope of this brief discussion. The etchants, in

SURFACE EXAMINATION

short, by selective chemical reaction, alter the polished surface to discolor it or corrode it. Characteristic coloration and/or corrosion effects such as pits and cleavages are diagnostic in analysis by the expert.

3. <u>Replication</u>: Replication is the process whereby the surface features of a specimen material are reproduced in "replacement" form from the original to a replicant material. The replicant is then examined microscopically. Objects that can be examined in either transmitted or reflected light are sometimes examined by replication to advantage. Machine parts and extended surfaces of many kinds, too large to be examined directly by the microscope, are especially suited to being examined by replication. Sampling methods involving dusts, particle distributions in the air and on surfaces are often collected by replicating materials which "fix" the distribution in place, allowing statistical studies to be made for industrial, health and hygiene, and pollution control purposes. There are two general ways in which replications are made: by film impression techniques and by gummed tape methods.

a. <u>Film impression</u>: Typical impression methods involve the use of a 4 percent solution of collodion in ether and alcohol. The solution is applied to the surface to be examined and then, when dry, "stripped off". The film retains even the finest structure by impressions of the surface sculpture. Polished metals, wood, skin, surfaces of plant leaves etc. are only a few such materials susceptible to this treatment. Examination of the film is done after mounting on a microslide, by light microscopy, either transmitted or reflected. Phase contrast can be used to great advantage in the examination of replication films, providing improved contrast for the finest details.

An alternate to the above is to apply the film to a microslide, then impress the specimen into it, pulling it free and leaving the impression. This technique is especially suited to the examination of fabric surfaces, fibers etc. In this type of replicating method cellulose acetate dissolved in ethyl lactate and amyl acetate is the medium often used. Normal cellulose lacquers are also used.

b. <u>Gummed tape</u>: Gummed tapes are sometimes used to replicate certain kinds of surfaces. On a plant leaf, for instance, the distribution of particles on that surface is easily studied by such means. The gummed tape is pressed onto the leaf surface and then removed and placed on a glass microslide for examination. If air currents are avoided during the impression of the tape, an accurate retention of particle distribution will be accomplished.

4. Semi-embedding: When transparent or nearly transparent materials are examined at their upper surface the undersurface image sometimes interferes with it. A method to reduce this effect is to semi-embed the material (fibers for instance) in a film of material that has a refractive index about the same as the specimen material itself. Thus, the influence of the lower half of the fiber is eliminated from the image of the upper (exposed) surface, enabling a more accurate assessment of its features to be made.

5. Casting: Inner surfaces of hollow specimens or voids within specimens can sometimes be examined, as mentioned before, by selective grinding and/or corroding methods. However, if complete inner surfaces require detailed examination in their entirety, those procedures are usually not satisfactory.

Microfossils, especially shells and/or tests of calcareous material can have their inner chamber surfaces and other details revealed by casting. A foraminiferal test, for instance, may be infiltrated with Canada balsam or other material. After the balsam is hardened, the test is then dissolved with weak acid to provide the shape of the chambers or cavities, and any connections between them, in the form of a casting.

6. Metallization: To improve the conditions for examining some surfaces microscopically, metallization is occasionally used. A more common method in electron microscopy, it can be used to advantage in light microscopy as well. These techniques are used to coat or evaporate metal onto the specimen surface. If, during the metallizing process the specimen is oriented (inclined) properly, metal deposition will "shadow" the surface. Some features under that condition receive a maximum deposition, some less and some not at all. Contrast is then a matter of surface topography which can be examined microscopically.

Even when metallization is done at normal incidence, more contrast is often evident by vertical illumination.

7. Artifacts: In all methods of specimen preparation for surface examination, artifacts may be developed. Often, the artifacts most often found in replicas are air bubbles, resulting from incomplete contact of the replicating material with the spec-

imen while making the imprint.

In grinding and polishing, and etching, surface features may be mechanically altered during the grinding or polishing process, and inaccuracies introduced thereby as to the true nature of the surface, or excessive chemical reaction may be very misleading to proper interpretation of results. Chemical etchants too strong, applied for an insufficient time or for too long a time will vary the coloration, pits and cracks to such a degree as to make comparisons with known results impossible.

Also, in many of the preparation methods of specimens for surface examination, especially replication, the interpretation must be made of an image that shows a negative aspect of a substitute surface. There is no better way to examine surface features than by direct examination of the original material. The substitution methods (by replication) must be carefully performed and interpreted to equal or approach the direct methods.

F. MOUNTANTS

1. Introduction: The reason(s) for using mounting media at all in specimen preparation may be either for enhancing visibility and perception of the object, or to improve its mechanical stability while under preparation or examination, or both. Further, the mounting media may be used either in temporary or permanent preparations. Temporary preparations may be for only the duration of the examination period under the microscope, or permanent preparations may be expected to last indefinitely.

The materials which provide an improvement in visibility and perception of the object are properly termed mountants. They also may provide the mechanical stability mentioned previously. When a material is used strictly for its mechanical properties it is perhaps more properly termed a cement. Between these extremes lie the immersion fluids which may provide either or both of these properties (optical and/or mechanical).

A colorless particle (object) can only be made visible with ordinary light microscopy if its refractive index differs from that of the surrounding medium. Contrast will be proportional to the difference between the refractive index of the medium and the object. It is for this reason that microscopical objects for the light microscope are commonly mounted in a medium of appreciably higher or lower refractive index than they themselves possess. If the microscope objective numerical aperture does not exceed 1.0, dry (air) mounting might be used rather than using a medium of higher refractive index (providing it is suitable for the object). However, depth of focus is decreased with dry mounts as compared with regular fluid or resin mounts. For instance, the apparent thickness of a layer of material is less than a similar layer of air in direct proportion to the indices of refraction. The axial magnification of a lens (the microscope objective, for instance) is equal to the square of the lateral magnification. If the lateral magnification is 4.0, the aerial image of a hemispherical object of 1.0 mm diameter will be a hemi-ellipsoid 4.0 mm in diameter and 16 mm high (or long). The value of a high index of refraction mounting media to minimize this effect is of value visually, becoming especially important in photomicrography.

Although a subject may contain structure within it on a resolvable scale, it may not be visible microscopically. This lack of visibility is frequently due to the inherent properties of the specimen material — namely that the structure possesses the same light transmitting or reflecting qualities as the surrounding or background medium. One of the foremost problems in light microscopy is the obtaining of contrast to reveal significant structure at high magnifications. The use of proper mountants in preparing the object for examination is an important step in solving this problem. If, for instance, the internal structure is to be studied, then it is sometimes an advantage to mount a nonhomogeneous object in a matching refractive index media. Choices of media to use depend greatly upon their refractive index and other properties.

2. Properties of mounting media

a. Refractive index: At least one of the most pertinent, if not the most important, properties of a mountant used in microscopy is that of its refractive index. The refractive index is a numerical expression of the deviation of a light ray (or wave front) upon passing from one medium to another. Referring to Figure 48, a light ray undergoes an abrupt change of direction upon passing obliquely from one medium to another wherein it travels with a different velocity. The ray will be deviated toward the normal (N) when the velocity is reduced and away from the normal when the velocity is increased. The relationship of the velocities and angles to the normal are:

$$\frac{\sin i}{\sin r} = \frac{v_1}{v_2} = n \quad \text{(refractive index)}$$

The ratio of the light velocities in the two mediums is a constant

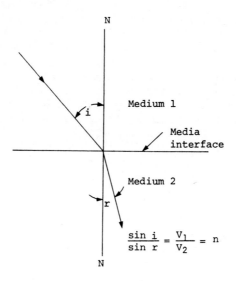

Figure 48. Refraction of a light ray.

for those media, and is called the refractive index of the second medium relative to the first. Refractive index figures expressed for microscopical mounting media (and most others as well) are compared with that of air, which is practically that of a vacuum and is assigned the numerical figure of 1.0. The physical meaning of the refractive index can be easily understood by considering a particular case. For the mountant Naphrax the refractive index (RI) is 1.70, meaning that the velocity of light in a vacuum (or air) is 1.70 times as great as in Naphrax, and also it will take the same time for light to travel through 1.70 mm of air as it takes to travel through 1 mm of Naphrax. Thus, the RI indicates the equivalent air distance of 1.0 mm of a substance.

The former indicates that light (of a given wavelength) travels at reduced velocities in substances other than air.

The refractive index of a substance (mountant) also varies somewhat with the wavelength of light. The velocity of an electromagnetic wave (lightwave) in any medium is

$$V = \frac{c}{\sqrt{K\mu}}$$

where K is the dielectric constant of the medium,
    μ is the permeability of the medium, and
    c is a constant 3×10^{10} cm/sec.

Although the value of K in the equation above is equal to unity for a vacuum, its value for other media depends upon wavelength. In data supplied with mounting media, the refractive index is referred to a particular wavelength of light (usually the "D" line of 589.3 nanometers — yellow light). Increasing the wavelength of the light from the yellow to the red decreases the refractive index and vice versa. For a given media, for instance, the refractive index might vary from a value of 1.5238 (F line — blue) to 1.5116 (C line — red). The temperature at which the refractive index is measured is also indicated, commonly being 20-25°C.

Because the peak of the visual brightness curve occurs not far from the D line, the index n_D has been chosen by optical designers as the basic index for ray tracing and for the specification of focal length. Two other indices, one on either side of n_D, are then chosen for purposes of achromatization. The most frequently used ones are n_C for the red end of the spectrum and n_F for the blue end (Figure 49). Refractive indices for mountants and

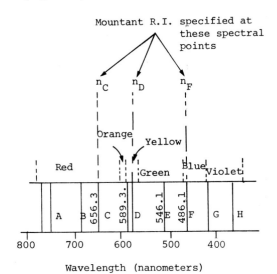

Figure 49. Refractive indices of mountants and Fraunhofer lines of the light spectrum.

immersion oils are most frequently expressed for these different Fraunhofer lines (D, C and F). Likewise, dispersion, an expression of the range of refractive indices for a given medium is generally limited between the F and C lines and indicated by $n_F - n_C$.

The quantitative effectiveness of differing refractive indices between object and mountant may be expressed by a so-called "visibility index". This is defined as a difference between the RI of the mountant and the object.

Referring to Table VIII, the "Index of Visibility" is shown for diatoms mounted in various media, indicating, for instance, that a diatom would be three times as "visible" in Hyrax as in Canada balsam. For example, (1.526-1.434) is 0.092 (index of visibility 9) and (1.700-1.434) is 0.266 (index of visibility approximately

TABLE VIII

Index of visibility

	Refractive index	Index of visibility
Diatom silica	1.434	—
Air	1.000	0.434
Water	1.334	0.100
Canada balsam	1.526	0.092
Styrax	1.580	0.146
Monobromide of Napthalene	1.658	0.224
Hyrax	1.700	0.266
Realgar	2.400	0.966

27) and, therefore, 27 divided by 9 is 3 (times the visibility of the one medium over the other).

However, visibility as such is dependent upon brightness as well as contrast and upon other factors too. Therefore, this type of comparison figure is a guide only to the effectiveness of increasing the RI of a mountant. Also, this index figure would seem to indicate that the higher the refractive index, the greater the visibility of the object. A very high index of refraction medium gives an object the appearance of being almost opaque, and for specimens of comparatively gross characteristics is not to be recommended. Extremely fine details, however, are enhanced by the use of very high refractive index media.

The commonly accepted dividing line between mountants of "high" and "low" refractive index is that of Canada balsam. Its longevity and popularity as a useful mountant has been largely due to its refractive index which nearly matches that of most glass. Its RI does vary considerably, however, from approximately 1.515 to 1.530. Mountants above and below approximately 1.52 are considered to be "high" and "low" index of refraction media, respectively.

b. Other properties: It is difficult, if not impossible, to make a general "line-up" of media and compare the merits of one against the other. Not only are there great numbers of them, but many are suitable only in specific application. Knowledge of the subject matter to be mounted, its characteristics and what is to be gained from microscopical observation will be determining factors in selection of media. Also, the matter of whether the mount is intended to be temporary or permanent will enter into a choice in many cases.

Therefore, it is the purpose in this section to elucidate properties in general that certainly will be considerations in choosing among the various media. When considering the various mountants, almost every physical and chemical property of them becomes a factor for one application or another. However, some properties may be deemed to be almost universally important considerations in all areas of work. They are:

(1) The refractive index.
(2) Chemical and physical activity.
(3) Neutrality.
(4) Penetrating qualities.
(5) Optical properties.
(6) Solubility.

The refractive index of media, its meaning and importance has been covered previously. A brief discussion of the remaining listed properties is also in order.

The chemical and physical stability of a mountant is very important, especially under long term conditions which pertain to permanent mounts. It should be resistant to crystallization and

MOUNTANTS

crazing or cracking when in a solid state (if that is the case), and not be hygroscopic, especially if a fluid media. For the latter, its vapor pressure should be nearly that of the average ambient water vapor conditions in the air. It should not oxidize over a period of time, and under conditions of preparation and storage be tolerant of a wide range of temperatures.

It should have a high degree of neutrality. It should be nonacidic and nonreactive with stains and noncorrosive or nondestructive of fixatives, cements and sealants used in mounting. Toxicity and disagreeable odors in a mountant are very undesirable. The preparation especially, and use of some types of special mountants are dangerous to health and life. Among these are mountants containing quantities of arsenic, or arsenic compounds.

A mountant ordinarily is used with specimen material that can and should be penetrated by it, to render it most effective in producing contrast and visibility of detail throughout the structure. In this respect, the mountants viscosity plays an important part. Low viscosity mountants penetrate better and do not entrain air to as great a degree as highly viscous materials. Some synthetics, or other mounting media, of higher molecular weight do not penetrate well (into limbs and vessels of insects, leafsection cell walls etc.).

Among the optical properties of a mountant that is important in almost any case is its color. With staining of a subject material for differentiation and contrast, a colored mountant might adversely effect the desired result. When the color of the object material itself is of diagnostic importance, the desirability of a colorless mountant is self evident. Optical activity such as polarization and/or fluorescence should be absent. Some small amount of the latter deficiency exists in many mountant materials but can be tolerated in ordinary work with the light microscope. Specialized light microscopy, such as that involved with petrography and fluorescence work, is very intolerant of such mountant defects, even to a minor degree.

Solubility of the mountant material in almost all areas is of considerable importance. The solvents to which the mountant material responds will quite often determine the applicability of it for a specific mounting task. Mountant solvents should be carefully determined and adhered to as crystallization and/or precipitation products in the final mount are quite often the result of improper attention to this detail. Sensitivity of the mountant to water is very important. Media that are soluble in water and those that accept objects directly from water are always attractive to the microscopical worker, as many procedural steps in mounting and numbers of reagents are thereby eliminated. On the other hand, many mountants (including synthetics) are very intolerant of water and precipitate, or crystallize in the presence of even minute quantities of it. Insufficient dehydration is often indicated by milky spots on the sections or clouds within the mounting media.

Where there is a choice of solvents for the particular mountant, one chosen with the lowest boiling point will insure minimum bubbles in the media. Always choose the least toxic one for safety. For instance, isopropyl alcohol may be used in many cases in place of xylene. It mixes with water, ethyl alcohol, chloroform, paraldehyde, glacial acetic acid, ether, oil of turpentine, acetone, xylene, melted hard paraffin, heated liquid paraffin and toluene. It dissolves Canada balsam, silicone DC804 and castor oil. Use of this substance instead of xylene in microscopical processing, where possible, very much minimizes the danger of poisoning by inhalation or through the skin. Other substitutes for the more dangerous solvents, if available, should always be used wherever possible.

The solubility of a mounting resin in ordinary ethyl alcohol is also very attractive as it might be expected to make it feasible to eliminate the usual clearing of tissue sections entirely. The mountant Pleurax, for instance, has this quality.

Since all resins used for mounting media show the highest refractive index when all volatile solvents are removed, any choice (if there is any) should include heating factors as a consideration. Lower heating rates may be mandatory of a particular specimen material in order to preserve its integrity, or no heating at all may be allowable.

In permanent microscopical mounts there are chemo-mechanical considerations of the mountant as well. Adhesion to glass or other substrates becomes an important factor. The time to harden (or dry) can also be a factor of consideration in both permanent and temporary preparations. Thermoplasticity is of considerable import with some types of preparations in that the mountant can be heated and reheated, being brought into and out of a liquid state indefinitely without chemical and/or physical changes in its properties. The Aroclors (chlorinated biphenyl and polyphenyl compounds) are examples of synthetic resins with this property. Some mountants have all of the desirable attributes except adhesion to glass, becoming very brittle upon hardening and detaching easily. A not commonly known fact is that opticians quite often add a plasticizer to Canada balsam or Dammar varnish

MOUNTANTS

used in cements for lens assemblies. A few drops of castor oil added to these substances is all that is necessary to improve their adhesion properties.

3. <u>Low index media</u>: Low index media are mostly nonresinous liquid substances, grading into jellied consistancies.

 a. <u>Aqueous media</u>: Mounting in water soluble media has always been attractive for the microscopist. Because many specimen materials do not withstand, without some distortion, the complete dehydration necessary for mounting in resins, mountants suitable for objects containing water have obvious advantages. However, most of the better known media with the latter characteristics are not suitable for stained preparations and so both object and media for this type of mountant must be chosen with care and possibly with some compromises.

The most widely used water-containing mounting media are preparations produced with glycerine or glycerine jelly. Glycerine jelly has a refractive index of approximately 1.414 and is suitable for mounting objects such as spores, pollen, insect wings, cornea of the human eye etc. There are various other low index media that have been developed over the years by private researchers. Polyvinyl alcohol has been one of the main constituents used, and chloral hydrate mixes have been very popular for the mounting of pollens and spores.

4. <u>List of mounting media</u>: Table IX lists mounting media by ascending refractive index. The list is not intended to be complete, but is representative of the wide range of substances used for mounting microscopical objects. In the remarks column comments are included that indicate some of the differences and peculiarities among the various mounting media. It is by no means a complete assessment of the mountant as would be required for application to specific uses. The source column indicates where complete information may be obtained and/or where such materials may be purchased. Also, it will be noted that considerable numbers of available immersion oils and fluids are not included in the list.

5. <u>Immersion fluids</u>: Mounting media of this category are liquid, usually not meant or suited for permanent mounts, and whose properties are used in analytical laboratory procedures with the light microscope, rather than to enhance contrast and visibility of the microscopical image.

TABLE IX
Mounting media

R.I.	Mountant	Remarks	Source
1.33	Pure water	Temporary mounts	3
1.384	CMC-10	Polyvinyl — alcohol mix/acidic, pH2 not recommended for tissue sections OK for pollen	3, 6
1.384	CMC-S	Same as CMC-10 with fuschin stain added	3, 6
1.411	Fluid mount 141	Inert — viscous — water white	3, 2
1.414	Glycerine jelly	Low heat melts/refrig. storage/pollen	3
1.426	Mount	1:1:1 glycerine, chloral hydrate, water animal feeds/food products	
1.455	Aquaresin	Viscous — water white — nondrying pH 8.0-8.4	5
1.47	Glycerol	Fluid mounts	3
1.47	Mineral oil		
1.47	Karo syrup		
1.483	Euparal	Nat. yellow/used w/analine dyes to prevent fading. Originally invented 1904 by G. Gilson	6
1.483	Euparal vert.	Green/for use w/hematoxylin stains solvent is euparal essence	6
1.483	Diaphane	Mounts directly from abs. alc. or 95% alc. colorless or green (copper salt). For use w/hematoxylin stains.	6
1.51	Histoclad	Neutral pH/low iodine content/120° M.P. Suitable for projection slides	1
1.515	Type VH imm. oil	Crystal rolling	2, 3
1.515 | 1.515 \} 1.525 |	Zeiss set	Non-hardening | hardening \} phase contrast hardening |	11
1.515-1.530	Canada balsam	Almost always acidic to a degree	6
1.52	Dammar balsam	Solvent xylene	6
1.540	H.S.R.	Manufactured by Hartman-Leddon Co.	7
1.540	Piccolyte		6
1.550	Caedax	Cyclohexanone. Neutral/water white solvent xylene/sensitive to H_2O	9
1.565-1.582	Polystyrene	Solvents: xylene, turpentine, oil of cloves, chloroform, oil of cinnamon, paraldehyde	
1.567	Permount	Solvent toluene, M.P. above 115°C, suitable for projection slides	4
1.58-1.60	Styrax	Natural resin — process lengthy /liquid amber orientalis/solvents chloroform, xylene toluene/diatoms	
1.569	Benzyl benzoate	Fluid mounts	
1.618	Balsam of Tolu	Natural resin similar to styrax	
1.62-1.65	MM165	Synthetic resin	2, 3
1.65	Hyrax	Napthalene derivative/solvents xylene, benzene, toluene, (Hanna)	1

continued

R.I.	Mountant	Remarks	Source
1.654	Aroclor 1260	Crystal rolling	3
1.660	Aroclor 5442	Particle mounting	3,5
1.66-1.70	Naphrax	Synthetic resin/sensitive to H_2O (Fleming) diatoms	2,8
1.65-2.1	Piperine and iodide mixtures		10
1.73	Aquaresin + potassium iodide		5
1.74	Methylene iodide	Nonresinous media	3
1.75-1.77	Pleurax	Sulfur-phenol resin/solvents 95% ethyl alc., isopropyl alc., acetone lemon yellow color (Hanna) diatoms	3
1.998-2.716	Sulfur-selenium mixtures	Toxic!	3,10
2.10	Phosphorus in carbon bisulfide	Deteriorates in a few years diatoms	
2.3-2.5	Realgar	7 pts arsenic - 23 pts. sulfur by wt. resinous — deep reddish brown color. Toxic! Diatoms/minerals	3,10
2.4-2.8	Thallium compounds	Thallium chloride/thallium bromide / thallium iodide/ Toxic!	10
2.72-3.17	Selenium and arsenic selenide mixtures	Minerals/Toxic!	10

Sources
Suppliers — Infromation — References

1. VWR Scientific
2. R. P. Cargille Laboratories Inc., 33 Village Park Rd., Cedar Grove, NJ 07009
3. Walter C. McCrone Associates, 2820 S. Michigan Avenue, Chicago, IL 60616
4. Fisher Scientific Company, P. O. Box 1307, 10700 Rockley Rd., Houston, TX 77001
5. Glyco Chemicals, Inc., 417 Fifth Avenue, New York, NY 10016
6. General Biological Supply House, Inc., 8200 S. Hoyne Avenue, Chicago, IL 60620
7. Arthur H. Thomas Company, Philadelphia, PA
8. National Biological Supply (NBS), England
9. E. Merck, Darmstadt, Germany
10. The Microscopic Determination of the Nonopaque Minerals, Second ed., Geol. Survey Bull. 848. Pages 11-17 incl.
11. Carl Zeiss, Inc., 444 Fifth Avenue, New York, NY 10018

a. Immersion oils: In Chapter I a brief discussion was included on the primary purpose and use of oils for immersion objectives. There are other uses to which they are put in microscopical work.

Their value for such other purposes is connected with optical and physical properties, the accuracy to which they are known, and the stability with which they are maintained. R. P. Cargille Laboratories, Inc., for instance, markets immersion oils of known qualities. Under these conditions some of the uses are:

(1) Temporary mounting media.

(2) Fluid mounting media. Particularly Type VH (very high viscosity) which is lighter in color than a number of other media and is stable. Permits rotation of mounted crystals and microfossils by slight shifting of the coverslip.

(3) Calibration liquid. The index of Cargille immersion oils is adjusted to 1.5150 with a tolerance of plus or minus 0.0002, making a reliable standard for calibration of many pieces of optical equipment.

(4) Optical coupling agent. Type VH, because of its high viscosity, has also been used both as a sealant and a coupling fluid, being applied between two optical elements.

(5) Transparency medium. Transparent materials and those with etched, round or translucent surfaces become transparent when immersed in an oil of the same refractive index. Since many glasses, plastics and fibers are approximately 1.5150, immersion oil has proven excellent as a transparency medium for many materials.

Optical and physical properties of Cargille Type A are listed in Table V. The same properties for other types of immersion oil can be obtained from the manufacturer.

b. Index of refraction liquids: These liquids are obtainable in sets whose refractive index ranges from 1.300 to 2.11. The classic use has been for identification purposes for ores and minerals. However, the use of them has become very widely applied, especially in recent years. Some applications in light microscopy include, but are not necessarily limited to:

(1) Identification of minerals, ores, chemicals, plastics and other transparent or translucent solids by immersion techniques.

(2) Index determination of synthetic fibers and other materials.

(3) Temporary mounting media to provide a large difference of refractive index between the media and specimen.

(4) Mounting and manipulating microfossils.

(5) Optical analyses and study utilizing techniques of dispersion staining.

(6) Particle identification work in air and water pollution investigations.

(7) Fragment identification in forensic microscopy.

Over the various ranges of refractive index liquids, the intervals in some sets are as little as 0.002 and different liquids might number more than 150.

Liquids in such sets may dissolve material which is immersed in them, or are very volatile and/or toxic. They also may be very sensitive to light, temperature and contamination. As a rule they should be stored in light-tight containers. Microscopists using these liquids should be well versed in their characteristics and application, not only for the preservation of their characteristics, but to avoid toxic effects in handling them.

6. <u>High dispersion liquids</u>. For dispersion staining techniques, the index of refraction liquids are required to have higher dispersions than the solids they are to match. Generally, commercially available high dispersion liquids cover the 1.500-1.800 refractive index range in intervals of 0.005. The optical constants for the F, D and the F, D and C (see Figure 49) lines are commonly supplied with them.

Cargille markets a Precision Dispersion Series (PD) that includes liquids made to order between refractive indices of 1.458 and 1.624 plus or minus 0.0002. The various index values are computed to the 5th decimal place for the h, g, F, e and C lines so accurate dispersion graphs can be drawn. This series is used where accurate dispersion curves are essential to determine refractive index after finding the matching wavelength on a spectrophotometer or other measuring equipment. A brief description of the use of such dispersion liquids is included in Chapter 4 and in detail in another volume of this series.

G. <u>MOUNTING METHODS</u>: After various treatments have been given to microscopical objects, they are almost always mounted on some sort of support which allows them to be examined with ease on the stage of the microscope. Because of the usual small size of such objects, they are mounted on supports large enough to be manipulated by the fingers and/or the mechanical stage. There are three general categories of specimen mounts: dry, resinous and fluid. Each will be briefly discussed.

1. <u>Dry mounts</u>: Dry mounts are considered to be those in which the specimen details to be examined are not immersed in any medium other than air.

a. <u>Box mounts</u>: This method of mounting is particularly suitable for rocks, minerals, seeds, small whole insects, crystals and microfossils. Many other types of objects may be profitably mounted in this manner as well. The specimen material is usually examined at low powers (10-40X) with incident light, and more often than not with a stereomicroscope.

Small paper, cardboard or plastic lidded boxes with dimensions of 1 x 1 x 7/8" are commonly used (Figure 50). The interior normally has a flat-black surface, either painted on, or by means of black paper inserts in the case of clear plastic boxes. Black opaque plastic boxes can be obtained that obviate the necessity to blacken the interior.

The specimen is mounted on a pedestal in the center of the box such that its top is just below the under surface of the lid. The pedestal is normally made to be of a smaller size than the specimen dimensions. This feature of construction allows for adequate incident illumination and makes the specimen appear to be suspended in a black void with the resulting contrast being both useful and pleasing, particularly with colorful and highly reflective material.

A pedestal is variously of wood (balsa is excellent), metal or cardboard. It is cemented to the bottom-center of the box with Duco or similar adhesives. The specimen itself may be fastened, depending upon its nature, to the top of the pedestal by similar adhesives. Water soluble materials such as gum tragacanth allow future adjustment and re-orientation of the specimen at will.

Variations of how the object is attached to the top of the pedestal allows considerably different types of material to be accommodated. Small paper, cardboard or metal discs of one-eighth to one-quarter inch diameter serve as platforms to support selected groups or microfossils, seeds, crystals etc., for comparative viewing. A straight pin inserted into the pedestal will provide a very (relatively) small support for individual crystals etc.

MOUNTING METHODS

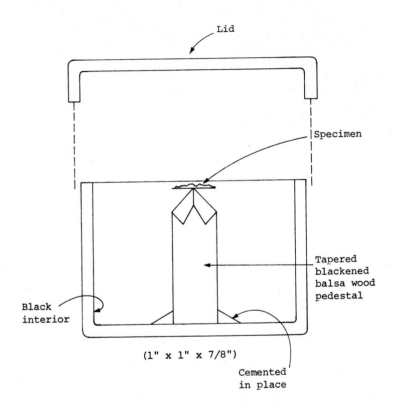

Figure 50. Typical box mount.

b. <u>Micropaleontological slides</u>: Special microslides are available commercially, or they can be made, for dry-mounting microfossils. In general, they are termed micropaleontological slides and commonly are of cardboard with dimensions of about 1 by 3". Their construction is usually that of two layers of cardboard, the bottom layer being thinner and of a continuous surface. The upper layer is thicker and is cut such that when cemented to the bottom piece it provides circular or rectangular "cells" into which the specimens are mounted. The bottom of the cell is usually blackened to increase contrast of the normally light colored specimen material. Viewing is from above by incident light at low powers.

Specimen material is usually fastened to the bottom of the cell using gum tragacanth which allows re-orientation of the specimen at any time by moistening the dried gum. Sable brushes of about 000 or 0000 sizes are used for picking, manipulating and orienting such specimens — usually under the stereomicroscope.

When the cells are small (circular, square or rectangular) they can be covered using ordinary coverslips to exclude dust. The covers are fastened to the top surface of the cardboard slide over the cavity with two or three spots of varnish or cement. The coverslip can then be easily removed with the edge of a knife or razor blade for re-orientation of the specimen etc.

Commercial micropaleontological slides of great convenience are composed of three parts. The cardboard slide with a cell or rectangular "well" (the latter often provided with a numbered grid in white ink on a black background), a cover consisting of a regular 1 x 3" microslide, and an aluminum clip which holds the assembly together. Refer to Figure 51.

    c. <u>Glass microslides</u>: Dry mounting is also accomplished using regular 1 x 3" glass microslides. This method allows objects to be observed either by transmitted light or incident light depending upon how the dry mount is constructed.

If the object requires a black (or colored) opaque background and is to be observed by incident illumination, a circular piece of paper of the required color or a painted background will often suffice.

A dry mount of this type requires that some sort of "cell" be provided to support the coverslip above the subject without crushing it. Such cells can be constructed in a number of different ways. Depending upon the thickness of the specimen to be mounted, a ring of varnish may be applied, using a slide-ringing turntable. The varnish is applied in layers, allowing drying between successive layers, until sufficient thickness has been built up. The coverslip is then applied to the last layer of varnish while it is still "tacky". Other types of cells are, of course, used. Cardboard, paper, plastic and glass have been and are used in

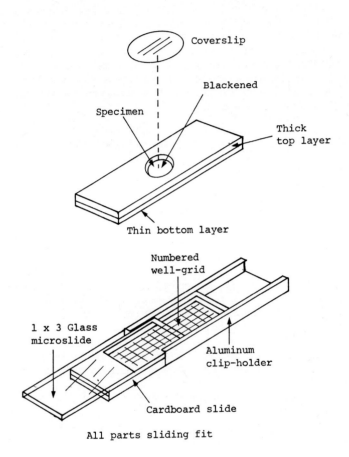

Figure 51. Micropaleontological slides.

such dry-mount construction. Glass rings of various thicknesses and diameters may be purchased for the purpose.

The major problem with dry mounts of the sort wherein they are sealed or semisealed is to maintain them in a really "dry" condition. Often such mounts become "fogged" on the undersurface of the protecting cover due to condensation of entrapped moisture. Careful preparation and thorough drying before the mount is covered will help prevent the condition. Objects suitable for such mounting include many of those which can be box-mounted, and also transparent or translucent material that can be mounted dry to advantage.

     d. <u>Polished specimens</u>: Because of their nature, polished specimens of metal, rocks, minerals and ceramics are usually mounted "dry". As they are examined by incident or reflected light the mount construction does not require transparent elements such as glass to be employed. More rugged materials, including metals, in keeping with the subject matter are commonly used. In Figure 52 a typical older type of mount for polished specimens is shown

The "cell" or container of the specimen was often made of a short piece of brass tubing — either circular or rectangular in crossection, brass being selected mostly for its workability. Other metals can be used if properly protected against oxidation and corrosion. The specimen was secured in a sealing wax matrix. Because sealing wax tends to flow under warm-weather conditions, two precautions were taken in such construction. One, the brass "cell" was peened with a punch to fix the sealing wax location, and the sealing wax was backed by plaster of Paris. Even with these precautions, this type of mount did not last indefinitely.

More permanent mounts of this type are made using Bakelite, molding it under pressure (up to 10,000 lbs/sq. in.) to form the matrix in which the specimen is held. Bakelite powders are available in black, red and green. The hydraulic presses used include built-in thermostated electrical heaters, and pressure gauges.

Because of the requirement for a press in making such moldings, and as more convenient materials have become available, the Bakelite method has given way, where possible, to cold mounting.

The cold mounting process is the modern counterpart of the sealing wax embedment described before. Exothermic acrylic

MOUNTING METHODS

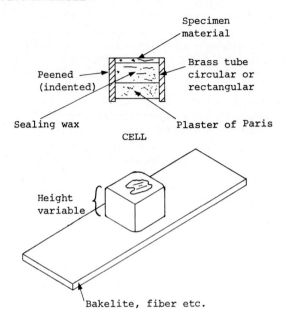

Figure 52. Polished specimen mount.

resins are now available that cure at room temperature. They ordinarily come in two parts — a powder and liquid. Proper proportions are mixed together and poured into aluminum rings in which the specimen has been placed. The rings are usually arranged on a glass plate. In about 30 minutes, specimens are ready for grinding and polishing.

Metallographic specimens mounted in this manner need protection against the highly polished surface being scratched and nicked, or corroded by fumes or moisture in the air. They should be stored in cabinets with protective drawer liners and with good circulation. Calcium chloride or some other desiccant should also be used to keep the storage atmosphere dry.

2. *Resinous mounts*: This category is by far the most familiar to the microscopist, and the great majority of microscopical preparations are mounted in this manner. The literature on microtechnique treats this method almost exclusively. The me-

chanics of making such preparations varies considerably with the mountant used and with the intent of the mounter. However, with all such mounts there are a number of common considerations that are worthy of, at least, brief coverage here.

Generally, this style of mounting encases the specimen material in a resin protected by a coverslip. Application of the resin, in its fluid form, its subsequent hardening and the placement of the coverslip often entails a number of technical difficulties that determine whether the mount is a good or bad one. In fact, at this stage of specimen preparation, all of the previous steps, sometimes very time-consuming ones, can be spoiled in their effect by improper technique.

a. <u>Penetration</u>: In many microscopical objects the quality of the final image is very dependent upon whether the mountant penetrated all voids and cavities in the specimen. As mentioned in a previous paragraph, the mountant with low viscosity, low atomic weight etc. will penetrate better.

However, other considerations such as refractive index etc. are involved and, with the selection of the mountant to be used fixed by those considerations, the technique followed is the only final factor left to the preparer assuring adequate penetration and permeation of the mountant throughout the specimen structure.

Judicious application of the mountant solvent is often the solution to proper penetration. If the specimen is moved through a stage of washing or flooding with the solvent of the mountant, it reduces the viscosity of the mixture to allow easy infiltration throughout the specimen structure. It is recommended that the mountant itself not be too fluidized by over-application of the solvent, but that it be retained in its normal state and the specimen and/or specimen site be treated with pure solvent before mountant application. This treatment will also improve conditions such that entrainment of air bubbles will be reduced to a minimum during penetration of the voids.

b. <u>Bubbles</u>: As mentioned previously, highly viscous mountants have the disturbing property of entraining air as they move into the voids and cavities of the specimen. However, high viscosity is not the only reason for bubbles to be present in a preparation. Sometimes the low boiling point of a mountant solvent is largely responsible for the generation of bubbles. Coupled with a viscous mountant, this can be very troublesome and most experienced workers arrive at procedures to minimize the effect. Quite often those procedures include very careful application of

MOUNTING METHODS

low heat, driving off much of the solvent prior to coverslip application, and proper application of the coverslip.

By using a vacuum pump one may sometimes offset the difficulty of bubble production of mountants being used because of their otherwise desirable properties. The preparation slide is enclosed inside an airtight enclosure in which the atmospheric pressure can be reduced; bubbles are literally "pulled out" of the mountant.

c. <u>Slide finishing</u>: After the preparation has been mounted in a resin, the coverslip applied, the mountant has been hardened by removal of its solvent and the slide generally cleaned up, a further step is often in order. That step is slide finishing and is especially important for permanent preparations.

Many mountants will react with air to oxidize or otherwise change chemically. For this reason, except where it is definitely not required, sealing the exposed edge of the mountant is important. If not done, after some time, the characteristics of the mountant may change. It may change in color — generally darkening, becoming cloudy; a precipitate may form or the mountant may even change its tenacity and separate from the slide or coverslip.

Sealants are available commercially for the purpose. They usually are applied with a small brush. Square or rectangular coverslips present a problem in making a neat-appearing seal. Round coverslips, in various diameters, are more suited to permanent preparations as sealants and other finishing touches can be applied with the aid of a "turntable" in evenly controlled layers to insure easy application and adequate protection. After the proper amount of sealant has been applied and it is dry, further work with a turntable to apply a layer of black or colored lacquer or paint as a decorative touch is easily accomplished (Figure 53). Material properly mounted in resin and sealed in the manner described are often, even after nearly a hundred years, as good as the day they were made.

Almost no cements or varnishes are even approximately waterproof as they will absorb from 3 to 10 percent of water if they are placed in a saturated atmosphere for a few days. However, with resin mounts, this tendency is minimal. With fluid mounts it becomes important and will be discussed later.

Cements made of drying oils harden by chemical change and sometimes are troublesome after long periods of time. Shellac sealants, on the other hand, harden simply by deposition from

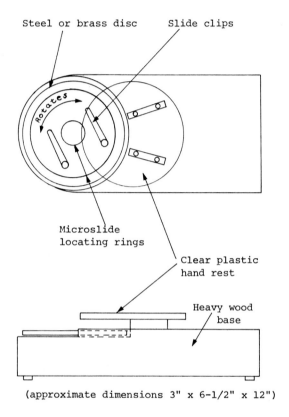

(approximate dimensions 3" x 6-1/2" x 12")

Figure 53. A ringing turntable.

solution, and exhibit long-lasting properties of stability, and are therefore preferable in most cases.

There are a number of slide ringing cements and finishing varnishes available commercially that are satisfactory. From various handbooks and laboratory manuals, formulas for compounding many different types of adhesives, cements, and varnishes are found that might be useful in various applications. A most useful material for sealing and ringing is gold-size. It is very tenaceous in its adhesion to glass and when dry is tough and pliable. It can be colored by the addition of lamp-black and used in the final finishing. Another very useful material for the pur-

MOUNTING METHODS

pose is white shellac. It also may be colored (if desired) for finish work. Glycerine jelly mounts can be effectively sealed with Dammar varnish or Venice turpentine.

3. Fluid mounting: Mounting specimen material in fluids for examination by the microscope is more common than supposed by most, and is done in a variety of ways. The major divisions of this type of specimen mounting are temporary mounts and permanent mounts.

a. Temporary mounts: Almost all types of materials are, or can be at one time or another, examined as temporary mounts, wherein after the isolated or at most a short period examination, the mount is discarded. Many cases of this involve both botanical and zoological specimens to examine them in the living state. The mounting media in those cases is often water or the fluids in which the specimens live. The mount is simply made by applying a drop of the fluid containing the organisms to be studied to a microslide and topping it with a coverslip. Evaporation is fairly rapid but if prolonged examination is necessary beyond the evaporation time, additional fluid may be applied at one edge of the coverslip with a pipette from time to time to replenish that lost (Figure 54). Also, if an absorbent material such

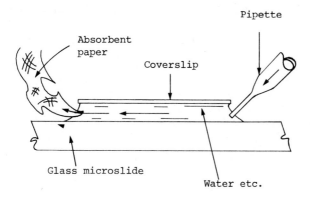

Figure 54. Fluid flow in a temporary mount.

as filter paper etc. is properly applied, the fluid beneath the coverslip can be withdrawn in controlled amounts. A corner of such material will cause, by capillary action, the fluid to move out from under the coverslip. The amount and rapidity of such withdrawal can be observed under the microscope and thus be controlled very easily. Likewise, the application of fluid at the opposite side of the coverslip can be observed microscopically. Observations involving the effects of fluid current action, replacement of fluid chemical consituents etc. are easily accomplished by this method, all while being observed with the microscope.

For prolonged examination of days or even of weeks duration, various experimenters have devised, from time to time, ways and means whereby the fluid mount becomes a permanent part of the microscopical setup. These methods provide for continuous replenishment of the fluid beneath the coverslip without the constant attention of the investigator. Some are quite simple in principle, and work well once adjusted properly. An example of an old but still useful scheme is shown in Figure 55.

Hanging drop slides are another type of temporary fluid mount. The method usually employs a microslide with a shallow depression in the center over which a coverslip is placed with a drop of the fluid to be examined. The coverslip is inverted over the depression and the fluid drop "hangs" within it. Because of the comparatively small enclosed space of the microslide cavity, evaporation will be slow and this type of mount will be comparatively long lasting. To further reduce the rate of evaporation it is sometimes convenient to add another drop of fluid or water to the microslide cavity.

A lying drop slide is another version of the latter. The microslide has a hole in its center, and a coverslip is cemented to its underside. The fluid to be examined is placed in the open hole and can be examined from the top (covered or uncovered) or from the bottom with an inverted microscope.

Many other types of fluid mounts have been devised over the years for culturing and/or prolonged examination of living organisms. Figure 56 illustrates a number of simple arrangements to examine living organisms under reasonably prolonged examination times. Some of these slides are cemented assemblies of ordinary microslides and specially shaped pieces of glass or other materials, and some are specially fabricated microslides with wells, slots, grooves, rings and other cavities trepanned or otherwise formed into them, requiring only the addition of a coverslip for completion of the assembly. All are designed to minimize fluid evaporation, and some provide for replenishment and /or aeration of the fluid during examination. "Moated" construction is also evident in some designs to prevent contact of the

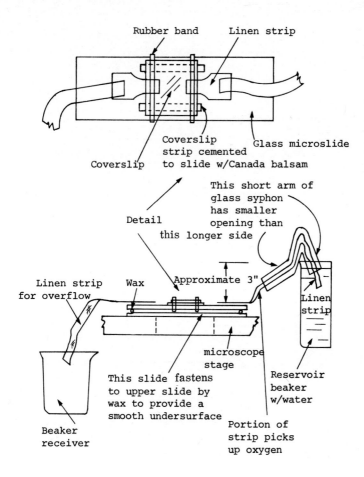

Figure 55. Constant flow temporary culture slide. (Circa 1890)

specimen-fluid with the temporary sealant used. Others, much more elaborate than those illustrated, provide for insertion or withdrawal of fluids and "atmospheres", and utilize wick or other reservoir systems for extremely long investigation periods.

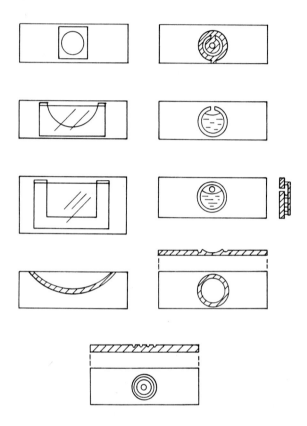

Figure 56. Temporary culture slides.

Microslides can be equipped with various types of electrodes and chambers for the application of electric currents to living microscopic objects. Most are not of the standard 1 x 3" dimensions, but are still small enough to be placed on the microscope stage. Figure 57 illustrates one such special application slide assembly. If strong electric currents and/or prolonged examinations are necessary, this type of slide needs to be equipped

MOUNTING METHODS

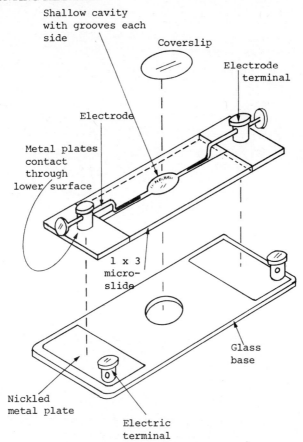

Figure 57. Electrified fluid mount.

with nonpolarizing electrodes. Such an electrode can be constructed as shown in Figure 58.

Simple fluid mounts for living cells mounted under a coverslip can be sealed by use of a silicone oil or silicone gum. Since these materials allow for the exchange of oxygen and carbon dioxide, such preparations can be maintained in a living state for a long time.

Zoological specimen material which is alive and normally examined with fluid mounts sometimes is in too rapid motion to be observed with the microscope. Various means are used to

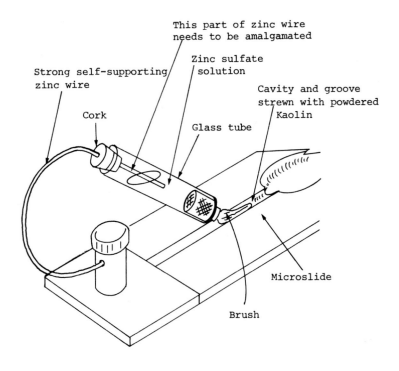

Figure 58. Nonpolarizing electrode.

slow down the rapid movements of microorganisms. Some entail narcotization of the animal (mentioned previously) or the insertion of an impediment in the fluid to restrict activity. Sometimes very small fragments of lens tissue, strands of cotton or glass wool can be used for the purpose. Other materials introduced into the fluid are gum tragacanth, gum arabic and very dilute solutions of fixing agents. All of these are successful to a degree but have disadvantages as well. The gums frequently distort shapes of the organisms and/or introduce impurities which are fatal. The fixing agents (1 percent formalin, for instance) will slow the motion, but eventually kill the living specimen. There are many other agents that can be used in this way as well, including commercial preparations designated for the purpose.

Allowing evaporation of the water beneath the coverslip to

MOUNTING METHODS

increase pressure on the living material will mechanically restrict movement. A specially constructed microslide, called a Rousellet Compressor (Figure 59), utilizes mechanical pressure

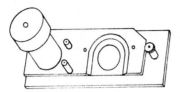

Figure 59. Rousellet Compressor.

to restrict microorganism movement. The fluid containing the specimen material is held in place between two glass plates mounted in a metal (usually brass) holder which is small enough to fit on a microscope stage. The distance between the glass plates can be varied by the action of a finely threaded screw. The top glass plate is usually of coverslip thickness. The space between the glass plates is decreased by the screw until the organisms are gently trapped, thus restricting their movements or immobilizing them for observation. Specimens as small as Euglena and Chlamydomonas can be examined without injury by this means. Since the coverslip and baseplate are always parallel, high power objectives can be used.

Another method for examining live material in fluid was described before the turn of the century in the journal of the New York Microscopical Society. The essence of the method is the use of a coverslip in a holder which can be slipped over the objective. With it in place the objective can be submerged into a comparatively large container (petri dish etc.) with specimens in fluid. Figure 60 illustrates the "immersion cap" in place on the objective and its uses. As a testimonial to the viability of many of the older techniques, this method is described in a reference that was published in 1973 (see end of chapter).

The cap can be fabricated to slip over the objective with a sliding fit or threaded fastenings can be made. It is important that the coverslip be sealed to the cap so that water or other fluid cannot enter. The slip-fit or threaded type of attachment is nec-

166   SPECIMEN PREPARATION AND OBSERVATION

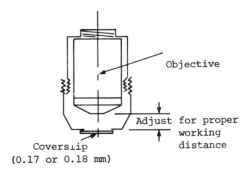

SCHEME FOR THREADED FASTENINGS

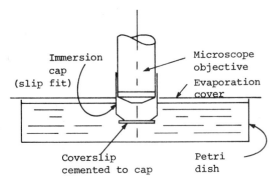

Figure 60. Immersion cap for fluid examinations. (drawing adapted courtesy Carl Zeiss, Inc.)

essary to adjust the coverslip distance to the front lens of the objective. It is recommended, for best resolution, that the water (or the fluid) level not be too far above the object. Because of the small amount of fluid present, a cover over the container is recommended to reduce the evaporation rate. A hole in its center, slightly larger than the diameter of the objective with immersion cap, provides access. The screw-type can be made so as to accommodate almost any objective diameter.

    b. <u>Permanent mounts</u>: Mounting specimen material permanently in a fluid media oftentimes provides a more lifelike appearance. Both animal and plant materials are suited to this type of mounting. Because of the difficulties of making such

MOUNTING METHODS

preparations, however, objects so mounted are usually restricted to those that are otherwise very sensitive to the various processes involved in resinous mounting. In order to give a reasonable transparency to the mounted object, the refractive index of fluid media is quite often above that of water but below that of Canada balsam. The media should be stable, resisting evaporation and not react chemically or physically with the sealing substance. In fact, it is these latter requirements that make fluid mounts of real permanence difficult of achievement. The fluid tends to leak out over a period of time after the sealant is applied, or vaporization occurs and gas bubbles appear, spoiling the specimen and/or its appearance.

The heart of the permanent fluid mount is the "cell". It must be of sufficient depth to contain the specimen material without pressure, and constructed with materials that do not react with the mounting fluid, even over long periods of time, and that in themselves are time-stable.

The fluids in which permanent mounts are made, of course, vary with the type of specimen material, and how it has been previously treated. One of the most useful and one which has withstood the test of time is a solution of 3 volumes of glycerine and 1 volume of water. Another that has been quite commonly used is lacto-phenol, consisting of 20 parts phenol, 20 parts water, 40 parts glycerine and 20 parts lactic acid.

A modern commercial product, "Fluid Mount", a viscous water-white nondrying material available from Cargille, is also very useful. Its refractive index is 1.411.

The cell in which the fluid mount is to be contained on the microslide can be constructed by building up successive layers of a sealant that is nonreactive with the fluid. This is done on a slide-ringing turntable and is quite suitable for shallow cells. Goldsize, bakelite, varnish and various products available commercially are used for the purpose.

When deeper cells are required than those that can be easily built up on a turntable, rings of metal or glass can be used. Sealing of them to the microslide and the applied coverslip is satisfactory using gold-size.

Soft-paraffin seals for fluid mounts have been made for years and are quite successful if proper technique is followed. References at the end of this chapter contain detailed procedures in fluid mounting.

4. Special considerations: In specific types of light micros-

copy some special considerations over and beyond the generalized ones are necessary. The inherent special optical conditions, illumination and other parameters demand careful treatment of the specimen in the preparation stage and in the selection of mounting materials and methods of observation. If attention is not paid to such details, a large part of the advantages of various types of microscopy will be negated or, at the very least, degraded.

As an example of some of the special considerations attendant to specific methods, the following notes are presented. They are by no means all-inclusive nor meant to be used as complete guidelines in special methods, but merely illustrative of the fact that special methods often require special precautions. Detailed methodology and techniques are covered in other volumes of this series.

    a. <u>Darkfield</u>:

        (1) At high magnifications the mountant or medium must not be too thick or concentrated or the required detail will be lost in the general glare and brightened background. The maximum allowable thickness for suspension of particles or organisms is about 10 μm.

        (2) For the same reason, microslides and coverslips must be scrupulously clean and free from scratches, and there must be no air bubbles in the immersion medium for these all show up brilliantly and decrease the contrast.

        (3) Where very critical work is contemplated, the effect of darkfield condensers can be enhanced by using microslides and coverslips made of quartz, as glass shows some fluorescence under the normally necessary intense illumination which decreases the background contrast. Also, quartz will better withstand the drastic cleaning methods necessary in such critical work. Of course, if quartz components become scratched with repeated use, better results can be obtained with new cleaned glass slides and coverslips. In any case, for best results slides and coverslips should be cleaned very thoroughly if absolutely clear darkfields are to be obtained.

    b. <u>Phase contrast</u>: The sensitivity of the method requires extra care in microtechnique. For best results, the following points apply:

        (1) Optically perfect slides and coverslip should be used. Slides and/or coverslips with nonplanar surfaces should not be used. Pits, cracks, striae, bubbles and

MOUNTING METHODS

unevenness will be detrimental to the final image as phase evidence of these will be starkly apparent, and interfere with features of the specimen image.

(2) Microslides and coverslips must be scrupulously cleaned. Any smudges, dirt particles, dust etc. will be detected and exhibited in the final image just as if they were imperfections in the glass.

(3) Specimen material should be bounded (above and below) by plane surfaces. If cavity slides and/or other substrates are used, the specimen image will be degraded or perhaps rendered useless.

(4) For maximum exploitation of the phase contrast method, choosing a suitable mounting medium is of the utmost importance. A properly selected mounting medium will insure satisfactory phase contrast and selective differentiation of structure in fixed specimens without stains. For both permanent and living specimen observation the mounting media with the proper RI can be found without difficulty, that with the optical properties of the specimen produces the best results. Special mounting media for the purpose are available commercially such that a wide range of choices is available to suit most specimen materials examined.

When there is a choice of observing modes, the following may be used as a guide:

(1) <u>Dark-low-low</u>. Since this mode of phase contrast is designed to furnish the strongest dark contrast having major differences in refractive indices, the subject is dark against a light background. Cells and living material are advantageously examined in this type of contrast.

(2) <u>Dark-medium</u>. This mode is best for image contrast for specimens with small phase differences such as fine fibers, granules and particles. Material is relatively dark against a light background.

(3) <u>Bright-medium</u>. Similar to darkfield, this contrast is especially suited for visual examination of bacteria, flagelli, protozoa, fibrin fibers, minute globules and for blood cell counting.

When crystals, plastics, fibers or small inclusions therein are to be examined by phase contrast, they should be well separated from one another within the specimen matrix, for example, in the form of separate entities in a homogeneous substance. In that case, a thin polished specimen or a thin section may be prepared. With this type of object material, attention to the difference between the refractive indices of the mounting medium and the specimen is more important than with biological subjects. The results will be dependent upon proper matching of the refractive index of the mountant to the specimen.

c. Interference microscopy

(1) Slides and coverslips should be of the best optical quality. They may not be wedge-shaped, have a wavy surface, striae or show optical strain.

(2) Avoid producing a wedge-shaped layer of mounting medium between slide and coverslip, and optical strain by applying pressure to the slide and coverslip.

If these precautions are not taken, continual readjustment of the microscope will be necessary every time the specimen position is changed. The adjustment would have to be made so as to maintain an unchanged interference background.

(3) Specimen material that is in the form of small units widely distributed over the field are better examined with a shearing type of instrument. If it is of such a size as to be as large as or greater than the field diameter, application of a two-beam interference system may be of greater advantage.

(4) If proper levelling adjustments are not made of opaque specimens being examined, the relationship between the observed fringes and the angle of surface tilt can introduce considerable error in measurement of fringe spacing.

d. Differential interference contrast: Since with this method of contrast, interference colors can be varied to suit any object detail from dark contrast to bright contrast, and there are some contrast color combinations that are more advantageous than others, the following is noted:

(1) When the background is dark, observation similar to darkfield is possible.

(2) When the background is gray, the visual sensitivity to detect a small path difference as a relief-like image becomes most pronounced.

(3) When the background color is magenta there is an ability to detect a path difference as a change in color.

MOUNTING METHODS 171

(4) When the analyzer is adjusted properly, brightfield observation is possible.

e. Polarization microscopy

(1) In thin section work with the polarizing microscope, particular attention necessarily is given to the thickness of the section. For insurance of adequate transparency and for identification of minerals, the section is usually standardized at 0.02-0.03 mm.

For other types of materials, especially wherein only structure may be of interest, thickness of sections may be as great as 0.25 mm or more.

Because of the nature of the work done with the polarization microscope and the importance of accurate measurements involving light intensity, patterns and colors, specimen thickness is probably more important than in many other disciplines.

(2) Whether the material to be examined is in thin section or in grains, it is frequently important that the refractive index of the mounting medium be known with considerable accuracy. In the majority of cases with grains it is more expeditious to mount them in oils.

Temperature effects on refractive index determinations are great and any exact statement regarding that characteristic must include the temperature at which it was determined. If temperature changes are accompanied by changes in chemical composition, optical properties may change as well.

(3) When examining platy minerals such as mica or other such materials in grain form, it is more difficult to manipulate them by "rolling" under the coverslip. It is sometimes helpful to mix a little powdered glass with the mineral or other grains to separate the coverslip and the microslide sufficiently so that the platy material can turn over.

f. Infrared microscopy

(1) In this microscopic method it is especially important that the samples or specimens sensitive to heat or photochemical effects be protected accordingly. This is often possible by choice of proper placement of the microscope relative to the source, monochromator and detector. If the microscope is placed between the slit of the monochromator and the detector, specimen material will be appropriately protected.

(2) Although the thickness of specimen material is usually about the same as for other methods of microscopy

(25 μm), there is a minimum area of sample required to provide sufficient energy at the detector. As there is an inverse relationship to source brightness in this regard, either the area must be sufficient at a given illumination level or the source intensity increased in the same proportion as the area of the sample is reduced. Therefore, sample size is of considerable importance with this method.

In addition, if the sample area is insufficient at the magnification used to completely cover the field, some radiation will reach the detector as "background" and effectively decrease the "radiation contrast" of the specimen under investigation.

(3) Impurity radiation from the specimen is also a factor of importance when quantitative work is being done. If this is not carefully controlled, published infrared spectra will be of limited value or of no use at all for analysis and identification.

g. Ultraviolet microscopy

When materials are to be examined by ultraviolet light the material quite often must be sectioned. When that is so, the embedding material must be chosen with care. It must either be transparent in the ultraviolet region of the spectrum or at least have a minimal absorption in the area of specimen interest. If there is a solvent required for the embedding material, the solvent itself must not absorb in the ultraviolet or it must be capable of being removed entirely. If not, then the solvent may either be retained in the embedding medium or absorbed by the specimen material, probably resulting in a false interpretation of the final image.

h. Fluorescence microscopy

(1) Because illumination levels of fluorescent radiation from specimen material are very low, special attention should be paid to the elimination of any stray light from any source that may interfere or mask it. The illuminator itself should be light-tight in this respect as nearly as possible, and also to reduce or eliminate harmful UV radiation.

Because the effectiveness of many fluorochromes are short lived under constant light exposure, some sort of light excluder should be incorporated at the source to minimize fading of the fluorescent specimen. This should always be in place when actual observation or photo-recording is not taking place.

(2) With all types of objectives used in fluorescence microscopy, the level of excitation can be increased by using

MOUNTING METHODS

immersion oil between the microslide and substage condenser. Of course, immersion oil is required for any darkfield system which requires an illuminating cone greater than 0.95 NA. Immersion oil for this purpose should not fluoresce. Oil that is known not to fluoresce, or glycerine, is recommended.

    5. Glass: A brief comment on glass used in the microscopical laboratory is in order. Although there are now literally thousands of variations of glass types by structure and constituency, there are two that are encountered most often.

One is soda-lime glass (SLG) which is basically the same as has been made for centuries, and the other is borosilicate glass (BG). Some glass qualities of importance and how these two compare are:

(1)	Optical clarity	(BG)	excellent;	(SLB)	good
(2)	Chemical inertness	"	"	"	poor
(3)	Structural integrity	"	"	"	fair
(4)	Warpage resistance	"	"	"	"
(5)	Leeching resistance	"	"	"	"
(6)	Effect on pH	"	"	"	"
(7)	Shelf life	"	"	"	"
(8)	Ease of disposal	"	"	"	excell.

It is seen that in almost all properties soda-lime glass is inferior to borosilicate. In microscopical work involving cultures it is evident that glassware should possess such qualities as to be inert, have a very high leeching resistance and have little or no effect on pH of solutions and specimen fluids. This is, of course, especially important with culture glassware, but of equal importance wherein microslides may be used in prolonged observation or culture work under the microscope.

Of some importance also is the quality of the glass structure. Microslides and particularly coverslips should be free of bubbles, striae, pits and cracks. Flat, parallel, plane surfaces are important and particularly so for certain types of microscopy as mentioned previously. The glass should be annealed and the edges of microslides of good quality are ground smooth to prevent chipping and crack development.

Some cavity slides are either not polished completely or polished poorly, showing ridges or irregularities in the cavity surface. They can produce very confusing images and should not be used.

Microslides of "noncorrosion" designation should always be

selected for any use, and of a thickness that is as close as possible to 1 mm. This is the most convenient thickness, being rugged yet thin enough for almost all optical purposes. Thick microslides should be avoided as they cause problems with condenser focusing. Very thin ones are easily broken.

Round coverslips of the proper thickness are recommended. They can be very easily and neatly sealed with a turntable. The most used size is probably 18 mm diameter. Diameters smaller (10 or 12 mm) are usually restricted in their use to selected or very small area specimens. The circular covers do not, in some cases, provide sufficient area. Rectangular coverslips of up to 24 x 60 mm are available. The latter are convenient for extended area specimens or serial sections or for dispersed specimen material in statistical studies.

H. SPECIAL OBSERVING CONDITIONS: Two types of conditions for the microscopist exist wherein mechanically specialized instruments are of great advantage. One is the condition where microscopic observation must be carried out under field conditions outside the laboratory. The other is specimen conditions that require radical departures from standardized instrumentation.

1. The portable microscope: The usual laboratory or research type of microscope stand is far too heavy and cumbersome to allow easy portability and use in any but laboratory conditions. This has long been recognized, and since before the turn of the century "portable" microscopes have been made by numerous manufacturers to supply the need. Most of the instruments offered at that time for field use were scaled down versions of larger stands, reducing weight in the interest of portability and to reduce the transport size in some cases, stands which could be disassembled, packed and reassembled in the field.

An example of an early version of a portable microscope was that produced by Bausch & Lomb in 1898. The microscope was similar to their "New" stand manufactured in 1897 except in size. It was smaller and was contained disassembled in a small leather covered carrying case. Assembly of this instrument provided that the limb be fastened to the carrying case acting as a base. The cover of the case could be raised or lowered to provide the inclination required for comfortable viewing. The essentials of the stand included a two objective nosepiece, a focusable substage condenser, substage mirror and focusing controls, but no coarse adjustment. The size of the case complete was 8 3/4 x 5 3/4 x

SPECIAL OBSERVING CONDITIONS

2 1/4" and its weight was 3 3/4 lbs. Other "travelling" microscopes of the time were made by Powell and Lealand, Swift, and Baker, all having their own peculiarities and attributes. Even into the 1930's and 40's manufacturers made "portable" microscopes by scaling down conventional stands and some are still available today in that form.

A rather radical departure from these designs provides a much smaller and lighter instrument, which can be hand held. A well known stand of this type is the McArthur Microscope devised by Dr. John McArthur of England. A precision instrument, capable of advanced work in the field, it is small enough to be easily carried in the hand or pocket. It has been used extensively in expedition work. One was taken for instance, on the 1954 Everest Expedition, and another across the Antarctic in 1957-8. Focusing is by fine adjustment only, objectives are parfocal and the viewing by them is done from below. Both achromatic and more highly corrected objectives are used, including oil immersion. Illumination is either external via a mirror, or built-in illumination is available with a miniature lamp powered by small dry cells. The entire instrument is about the size of a 35 mm camera. Improvements in its operation and performance continue to be made, and the McArthur Portable Microscope has become a standard in the field.

In addition to the stand described above Dr. McArthur designed a simpler version of this type of instrument for the Open University in England. In general appearance and size it is similar, and its capabilities are considerably reduced. However, it is a very convenient little instrument of very light weight. The case is of high-impact plastic of compact design, dimensions being approximately 3 x 5 x 1". Figure 61 illustrates the construction features.

The focusing wheel alters the position of the objective lens, but not that of the ocular. The image in this microscope is erect but reversed from left to right. The objectives are changed by moving a detented slide. The illumination may be changed from external to internal by moving the metal mirror. Optics include a 10X Hugenian ocular and 8X and 20X objectives. With the 20X objective resolution of 1 μm is possible. Although designed primarily for educational use by individual students, this little instrument serves admirably as a field microscope of considerable ability and convenience.

Another but much more sophisticated version of a portable microscope is available from Nikon. Figure 62 illustrates this

176   SPECIMEN PREPARATION AND OBSERVATION

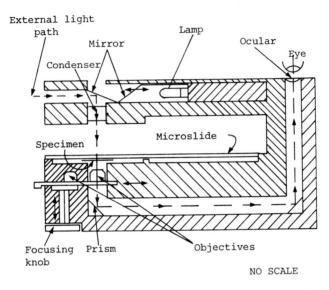

Figure 61. Essentials of McArthur Open University Microscope.

Figure 62. Precision portable microscope. (photo courtesy Nikon.)

compact precision instrument. Practically every convenience found on a conventional research stand is offered here. This style of portable microscope may be obtained even with phase contrast objectives, turret condenser and centering telescope. Polarizing accessories are available and other accessories especially designed for use with it, including a blood counting chamber and an objective for uncovered microslides.

The light weight and size of these smaller portable instruments allows them to be carried anywhere on the person just as a small camera is. Instant observation at streamside, lakeside or any location can be performed by holding it in the hand and directing it at the sunlit sky for illumination. Provisions are often made for internally supplied illumination and a means for mounting the instrument on a camera tripod.

With the advent of a considerable increase in environmental research, more and more emphasis is being placed on microscopy in the field. For instance, McCrone Associates of Chicago not only markets a pocket microscope by Tiyoda of Japan, but also features a complete sample preparation field kit providing facilities for collecting, documenting, microscopical checking and morphological analyses on the spot.

2. The inverted microscope: An example of microscope design being adapted to special specimen conditions is the inverted microscope. This style of instrument is also not new by any means as one of the first was made in the 1880's by the Grunow brothers in the United States based on a design by Professor J. Lawrence Smith of Kenyon College in Gambia, Ohio. The primary function at the time was observation of chemical reactions without fumes affecting the objectives or rising directly into the microscopist's face. Of course, advantages in observation of biological, medical and other specimen material became obvious soon thereafter.

The inverted microscope has become an important member of the community of microscopes, and today is used in diverse fields. Applications include metallurgy, chemistry, geology and biology. The instrument has been found to be admirably suited to tissue culture work wherein specialized chambers, flasks or petri dishes are used. In fact, this particular area of research has initiated design features and accessories specifically suited to its purposes. Stage assemblies, for instance, are designed to eliminate drifting during long-period, time-lapse cinemicrography, and numerous types of holders are available for the special-

178 SPECIMEN PREPARATION AND OBSERVATION

ized glassware used on the stage. In some work, wherein micromanipulators may be used, it is desirable that the stage level remain fixed. Inverted microscopes can be obtained with either the stage being focused or the stage remaining fixed with the optics being focused.

The inverted microscope today is available with almost any option found with upright instruments, including photomicrography, phase contrast, episcopic or transmitted illumination, and batteries of optics including specialized long-working-distance objectives, polarizing attachments, rectangular and circular mechanical, and glide stages. So diversified have they become, it is not possible in a limited space to completely cover all a aspects of their application, operation and equipage.

Figure 63 illustrates a Nikon model set up on a vibration-isolation table, with provisions for transmitted light and photomicrography. An electrically timed shutter control box is to the left, and the transformer and voltage control for the illumination to the right.

For metallurgical work, the inverted microscope is commonly equipped with a reflected light illuminator, allowing the examination of polished sections of opaque specimens with the use of all common microscopical techniques.

Figure 63. Inverted microscope on vibration isolation table. (photo courtesy Nikon).

I. REFERENCES AND COMMENTARY

1. Evens, E. D., "Some notes on glycerine mounting media," JQMC (Ser. 4), 5, No. 16 (1960).

2. Barnett, W. F., "A method of mounting in fluid media using synthetic resins," JQMC, 29, No. 4 (November 1962).

3. Dade, H. A., "On mounting in fluid media, with special reference to lactophenol," JQMC (Ser. 4), 5, No. 11 (August 1960).

4. de Ternant, P, "Some uncomputed factors in the problem of permanent fluid mounting," JQMC (Ser. 3), 1, No. 4 (October 1935).

5. Crossman, G. C., "The use of silicones as sealants for living cell preparation," The Microscope and Crystal Front, 15, No. 9 (March-June 1967).

6. Phoenix, Eric. A.,"Experiments in making fluid mounts," The Microscope, 12, No.3 (November-December 1958).

7. Frison, Ed, Translated by D. S. Spence, "E. Kaiser's glycerol-gelatine," The Microscope, 10, No. 7 (May-June 1955).

8. Salmon, J. T., "A new polyvinyl alcohol mounting medium," The Microscope, 10, No. 3 (September-October 1954).

9. Crossman, G. C., "Mounting media for living phase microscope specimens," The Microscope, 10, No. 1 (May-June 1954).

10. Robinson, A. I., "Canada balsam," The Microscope, 2, No. 6 (July 1938).
Describes physical and chemical characteristics of the mountant in detail (RI, acidity, optical activity, fluorescence etc.).

11. Anon., "Temporarily mounting metallurgical and rock specimens parallel to the stage," The Microscope, 1, No. 2 (September 1937).

12. Micron., "Gold size and its properties," The Microscope, 2, No. 1, 5-6 (February 1938).

13. Dade, H. A., "A new sealing method for fluid mounts," The Microscope, 2, No. 1, 14-15 (February 1938).

14. Larsen, E. S., H. Berman, "The microscopic determination of the nonopaque minerals," Geological Survey Bulletin 848 (1934, reprint 1964).
Contains many facts about and procedures for the preparation of very high refractive index mountants and liquids.

15. Evens, E. D., "Styrax," JQMC (Ser. 4), 5, No. 9 (February 1960).

16. Editor, "Polyvinyl alcohol mountants," The Microscope, 9, No. 1 (May-June 1952).

17. Frison, Ed, "Some further experiments with synthetic resins as mounting media of high and low refractive indices," The Microscope, 10, No. 8 (July-August 1955).

18. Frison, Ed, "Courmarone resin as a mounting medium," Part I, The Microscope, 9, No. 2 (July-August 1952); Part II, Ibid. 9, No. 3 (September-October 1952).

19. Scott, T. L., "Wanted, an ideal mounting medium," The Microscope 8, No. 8 (July-August 1951).

20. Salmon, J. T., "Polyvinyl alcohol as a mounting medium in microscopy," The Microscope, 8, No. 6 (March-April 1951).
A discussion of various preparations of this low RI medium, and applications in mounting insects and similar specimens.

21. Scott, T. L., "Gum chloral - a mounting medium," The Microscope, 5, No. 5, 122-125 (March 1943).
Preparation and use of the medium is discussed.

22. Sylvester, G. R., "The effects of mountants on the permanence of staining differentiation in haemotology," The Microscope, 5, No. 6 (1943).

The results of a ten-year test on slides mounted in neutral Canada balsam and neutral Euparal, with and without Haden's solution.

23. Short, M. N., "Microscopical determination of ore minerals," Geological Survey Bulletin, 914 (1940).

24. Delly, John Gustav, "Mounting media for particle identification," The Particle Analyst, 1, No. 8 (April 1968).

25. Allen, R. M., The Microscope, D Van Nostrand Co. Inc., Fifth Printing, 1947.
A good elementary introduction to the general field of light microscopy. Many practical aspects of mounting microscopical objects are included.

26. Gray, Peter, Handbook of Basic Microtechnique, Third edition, McGraw-Hill Book Company, 1964.
A supplement to Gray's "Microtomists' Formulary and Guide," and one of a series in McGraw-Hill publications in the biological sciences.

27. Gray, Peter, Gray's Microtomists' Formulary and Guide, McGraw-Hill Book Company, 1954.
A standard reference for the microscopist in technique and methods.

28. Barnett, W. J., Freshwater Microscopy, Constable and Company, Ltd., 2nd ed. 1965.
An introductory treatment of pond life and the microscopical techniques used.

29. Clark, G. L. (Editor), The Encyclopedia of Microscopy, Van Nostrand Rheinhold Company, 1961
Appropriate sections treating topics included in this chapter.

30. Hanna, G. D., L. A. Penn and P. Ruedrich, "Another synthetic resin useful in microscopy," Science, 70, 16-17 (1929)

31. Hanna, G. D., "Hyrax, a new mounting medium for diatoms," J. Roy. Micro. Soc., 50, 424-6 (1930).

32. Cameron, E. N., "Notes on the synthetic resin hyrax,"

American Minerol., 19, 375-83 (1934).

33. Fleming, W. D., "Synthetic mounting medium of high refractive index," JRMS, 64, 34-37 (1943).

34. Fleming, W. D., "Naphrax, a synthetic mounting medmedium of high refractive index; new and improved methods of preparation," JRMS, 74, Part 1 (June 1954).

35. Hanna, G. D.,"A synthetic resin that has unusual properties," JRMS, 69, 25-8 (1949).

36. Anon., "The acidity of Canada balsam," Quart. J. Pharm. Pharmacol. 11, 709-13 (1938).

37. Schlueter, G. E., "The inverted biological microscope — origins to a modern design," Am. Lab., 4, 62 (1972).

38. Jedwab, J., M. Defleur-Schenus and A. Herbosch, "Mounting and polishing small quantities of minerals," Am. Mineral., 55, 1065-6 (1970).

39. Senior, W., "Polyester resin as a mounting medium for light microscopy," J. Microscopy, 91, 207-10 (1970).

40. Heunert, Hans-Henning, "Microtechnique for the observation of living organisms," Zeiss Information, 20, No. 81 (1972-73).

41. Heunert, Hans-Henning, "The many different uses of the inverted microscope," Zeiss information, 71 (1969).

CHAPTER 4

SAMPLE CHARACTERIZATION

A. INTRODUCTION: Much information regarding a microscopical specimen is obtained by simple visual examination with the microscope. However, usefulness of the light microscope goes beyond the mere resolution of detail. The complete knowledge of any subject extends to its hidden as well as its obvious characteristics. This chapter will briefly examine various methods which are used in light microscopy to extend our awareness of sample characteristics. Many factors characteristic of a material, specimen or object may be revealed by the microscope either by special methodology or auxiliary apparatus, or both. Because of the nature of the instrument itself, most of the characteristics revealed will be by the action of light in the visual range of the spectrum. Others included are in the nature of specific shape and form, dimensions, weight, and in electrical, magnetic and chemical reactions.

Individual facts gained as a result of microscopical examination in this manner constitute a characteristic of a material or specimen, sometimes unique, sometimes not. A collection of many or all of the characteristics of the object usually constitutes a unique characterization of it. This type of procedure is very important in any analysis with the microscope and particularly important in identification work.

B. CHARACTERIZATION BY LIGHT PHENOMENA: Light behavior is revealed to us in the form of reflection, refraction, dispersion, diffraction, interference, absorption, polarization and excitation. A brief description of each of these phenomena will provide a basis for understanding the various specimen characterization methods by light action with the microscope (Figure 64).

    1. Reflection: Reflection of light from a smooth surface, like that of a mirror, takes place along a definite direction determined by the direction of the incident ray, and is called regular or specular reflection.

Reflection from a rough or matte surface occurs in a great many directions for any one direction of the incident beam and is said to be diffuse or scattered. It is by diffuse reflection that nonluminous objects become visible.

    2. Refraction: Reference is made to Figure 48 in Chapter III. This figure indicates that a ray of light of a single wave-

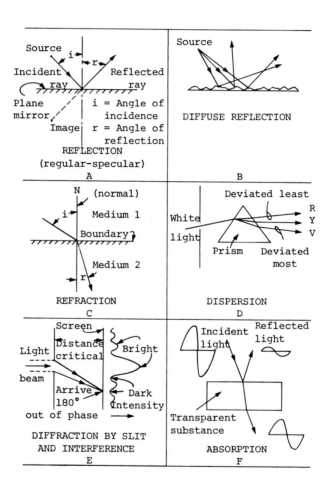

Figure 64. Some light characterization phenomena.

CHARACTERIZATION BY LIGHT PHENOMENA 185

length undergoes an abrupt change of direction upon passing obliquely from one medium to another. The ensuing brief discussion in Chapter III in relation to the figure is appropriate here.

A practical application of refraction is the Becke-line test for refractive index illustrated in Figure 65. It will be noted that

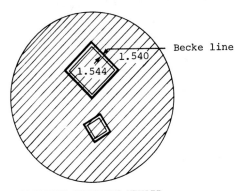

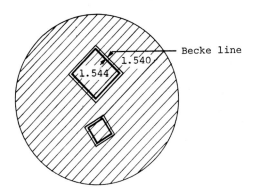

Figure 65. Characterization by refractive index.

the bright line (Becke line) moves toward the material of higher refractive index (in this case the sodium chloride crystal) upon focusing upward with the microscope, and in the opposite direction when focusing downward. Use of fluids of known refractive index in a bracketing manner will determine the refractive index of the solid.

3. <u>Dispersion</u>: The spreading out of a light beam into its component colors is known as dispersion. For instance, when a narrow slit is illuminated by white light which is then passed through a prism, refraction takes place and the constituent colors are spread out into an array called a spectrum. The longer wavelength components are deviated the least and the shorter wavelength components the most. The type of material through which the light is dispersed determines the deviation of any particular wavelength, different materials having different properties of dispersion. The deviation is a measure of change of direction for a particular wavelength of light in a particular medium.

4. <u>Diffraction</u>: Although it is commonly said that light travels in straight lines, careful observation shows that it bends slightly around the edges of an obstruction. The spreading of a beam of light into the region behind an obstacle is known as diffraction. A parallel beam of light of a single wavelength, after passing through a slit, will ordinarily produce a bright band somewhat wider than the slit; this band will be bordered at the edges with a few narrower bands which are alternately dark and light.

A circular hole will produce a round patch of light surrounded by a few rings which are alternately dark and light. The light and dark portions of the images for each case are the result of constructive (in phase) and destructive (out of phase) interference portions of the light wavefront propagating through the opening. The size of the opening in terms of the wavelength of light involved and the distance from the slit or hole to the incident surface upon which the interference is detectable will determine the intensity and distribution of the pattern.

5. <u>Interference</u>: The superposition of two light waves upon arriving simultaneously at a given point will produce a total illumination that depends upon their wavelengths, amplitudes and the phase of one relative to the other. Diffraction, as previously explained, provides a separation of light components from a single source at a single wavelength that recombine in the manner described to form variations of light intensity of a single frequency or wavelength which is dependent upon phase and amplitude only.

CHARACTERIZATION BY LIGHT PHENOMENA 187

The third factor of frequency, or wavelength, differing in the two original light sources will further create color differences in the total illumination pattern.

With white light of a single source, in which many colors are blended, the annulment of one color at a particular point still leaves illumination by the other colors. When two or more waves of a single frequency are in phase the resultant is maximum intensity and when they are out of phase, or in phase opposition, the resultant is minimum amplitude or intensity.

The conditions which produce interfering light waves are numerous; those produced by diffraction of light passing through very small slits, orifices, structural details etc. or around edges of structures, have been briefly described. Lenses also produce diffraction effects.

Colors due to the interference of light waves at the front and back surfaces of very thin films are observed by reflection. "Newton's rings" are a special effect of this type, being the result of light interference from that reflected by both surfaces of the air film between the convex and plane surfaces.

Some natural substances have the property of double refraction, such as quartz, calcite and mica. Man-made devices such as diffraction gratings and others are often employed to produce interference effects.

The phenomena of interference, like most of the other properties of light, are used in microscopy both as a tool to investigate object materials, or the phenomena produced naturally by the object material are interpreted as a characteristic.

6. <u>Absorption</u>: The amount of light that is reflected from a transparent substance depends on the angle of incidence and the refractive index of the substance. At perpendicular incidence of a light beam only a part is reflected, the rest being absorbed and transmitted.

The amount of absorption of light depends on the nature of the absorbing substance and its thickness. Physically, absorption means the conversion of electromagnetic energy to heat energy.

7. <u>Polarization</u>: Light vibrations restricted to a single plane are said to be plane-polarized. The brief treatment of this phenomena in Chapter III under polarizing filters is appropriate to the discussion here. As in the other light phenomena, polarized light is used as a tool or is interpreted as an object characteristic.

Light is not only polarized in a single plane as mentioned

188   SAMPLE CHARACTERIZATION

above, but is also polarized "circularly". Circular polarization can occur naturally or can be effected purposefully to obtain certain effects in microscopy. Because it is used in a number of microscopical equipments and devices, it is important enough to warrant a brief explanation of how it can occur.

The sequence of events examined is clearly depicted in Figure 66 along the oblique line denoted by arrow heads traversing the figure from left to right. At the upper left a vector diagram

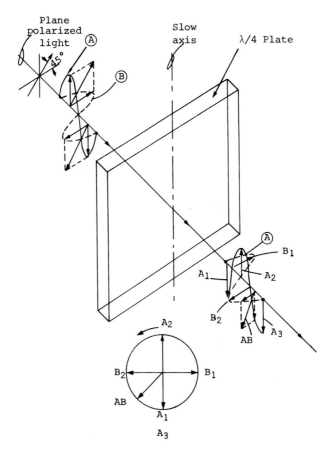

Figure 66. Circular polarization by quarter-wave plate.

CHARACTERIZATION BY LIGHT PHENOMENA 189

(in a plane at right angles to the direction of propagation) indicates plane polarized light at an angle of 45° to the horizontal. For any one particular frequency, the amplitude (length) of the vectors will vary from zero to maximum and back to zero in a period determined by that frequency.

Assuming a sine wave variation in amplitude, the 45° vector can be generated by two vectors varying at the same rate and amplitude. This is indicated by two sine waves (A and B), one in the horizontal plane and the other in a vertical plane, which at any one instant in time provide the resultant vibration at 45° to the horizontal.

If a substance of a specific characteristic, of a thickness of one-quarter wavelength (or an odd multiple thereof) is interposed in the path of propagation of the plane polarized wave, an interesting action takes place. Certain crystalline substances (such as mica) have the property of having differing refractive indices in different planes. If a specially cut piece of such a substance is oriented so that the so-called "slow-axis" is vertical, the sine wave component that is vertical in the figure will be "slowed" or phase retarded by one-quarter wavelength (90°). The other sine wave component in the horizontal plane will not undergo this relative quarter-wave "delay". The resulting sine waves are indicated as emerging from the quarter-wave plate along with spatial vectors indicating the resultant vibration from their combination. It will be noted that sine wave A (the vertical one) is shown as being phase retarded by 90° from that of the sine wave B (the horizontal one). Close examination of the two sine waves and the vectors representing instantaneous amplitude values shows that the resultant vector (combination of the individual vectors of sine wave A and B) does not increase and decrease in amplitude (from zero to maximum and back to zero) in a period of time determined by the frequency, as it did with a plane polarized wave previously. Vectors A_1, A_2, A_3, B_1, B_2 indicate maximum amplitudes of their respective sine waves when the other component sine wave amplitude is zero. At any other time, wherein neither sine wave amplitude is zero such as at AB, the resultant vector of the two is always the same amplitude as the maximum amplitude of either of the two, but at a different spatial position relative to the horizontal and vertical. The result is that during a time period corresponding to one cycle of amplitude variation (one wavelength), the vector result is a constant amplitude vector that rotates once in a plane that is normal to the direction of propagation. This is shown graphically by the circular figure denoting the various vector positions in space. At time zero, the vector amplitude is maximum and downward (A_1); 90° of the cycle later (one quarter wavelength), it has the same amplitude but is now to the right, and so on. This result, in which the amplitude of the light wave remains constant but its plane of polarization changes at a rate governed by its frequency (wavelength) is termed "circular polarization". In this particular example, the circular polarization is in a counterclockwise direction (as viewed back along the line of propagation). If the quarter-wave plate had been positioned such that its "slow axis" were horizontal instead of vertical, the sine wave component B would have experienced the 90° phase retardation, and the circular polarization would have been clockwise in its rotation.

The quarter-wave plate, therefore, produces a quarter wave or 90° phase change, and circular polarization, either clockwise or counterclockwise dependent upon the orientation of the "slow axis".

A half-wave plate produces a 180° phase change, but the output is plane polarized and the plane of polarization is rotated with respect to the input plane, the amount being 2 times the angle between the input plane and plate axis.

A "rotator" plate produces fixed values of pure rotation of the plane of polarization. The value is independent of plate orientation and only dependent upon wavelength and plate thickness.

8. <u>Excitation</u>: Certain substances when exposed to or excited by light continue to emit light a fraction of a second after removal of the exciting source. This property of a substance is known as fluorescence and is generally defined as: luminescence stimulated by radiation, not continuing more than about 10^{-8} second after the stimulating radiation is cut off.

In general, the light emitted by an excited object is of longer wavelength than to which it is exposed. Briefly, the incident light excites the atoms of the particular substance, and the longer wavelength of the emitted light is attributed to the return of the displaced electrons to their original energy levels in two or more stages rather than in a single transition.

Since certain specific substances are affected in this way, producing fluorescence of various magnitudes and colors (wavelengths), fluorescence phenomena have become a very powerful analytical tool in light microscopy.

C. <u>POLARIZED LIGHT MICROSCOPY</u>: The use of polarized light was briefly discussed in a previous chapter relating to the

POLARIZED LIGHT MICROSCOPY

improvement of contrast in the microscope image. That use is limited and of rather minor importance compared to the extensive quantitative as well as qualitative work that may be done using this property of light with the microscope. The following will provide an outline of the various characterizations that can be obtained using polarized light techniques.

1. The polarizing microscope: To characterize materials on a microscopic scale requires more than the polarizer and analyzer accessories previously discussed for use in the enhancement of contrast. A brief description of the salient features of such a specialized instrument is appropriate at this time (Figure 67).

The ocular is usually equipped with cross hairs. Below the ocular in the bodytube of the instrument is the Bertrand lens. It is arranged such that it can be swung or slid in and out of the axis of the tube, and sometimes has an adjustable iris diaphram just above it. It is used in examining the back focal plane of the objective. Below this but above the objective in the tube of the instrument is the analyzer. Provision is made for its rotation through at least 90°. Between the analyzer and objective is an accessory slot into which other devices such as a quartz wedge for analytical work may be placed.

The nosepiece carrying the objective is provided with centering adjustments, or the objective mount itself is so equipped.

For quantitative and exacting work the objectives are "strain-free" versions, with glass-to-air surfaces antireflection coated. Strain-free design precludes the setting up of mechanical stresses in the glass elements by special glass selection and/or construction of lens mounts. Sometimes they are merely especially tested and selected from a manufacturing lot to assure the strain-free condition. The antireflection films enhance image contrast and reduce possible disturbing interference patterns at the glass-air interfaces.

The stage is a rotating one which is graduated in degrees. A vernier is included to permit reading of angles to the nearest tenth degree.

The substage contains the condenser, iris diaphram and carrier for the polarizer, and sometimes an additional carrier for a sensitive tint plate.

The preceding facilities are rather minimal and their arrangement, as described, more or less typical. More elaborate stands with slightly different mechanical designs are, of course,

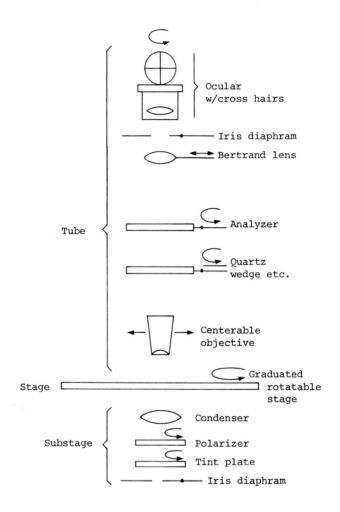

Figure 67. Polarizing microscope essentials.

POLARIZED LIGHT MICROSCOPY

in common use.

2. <u>Characteristics determined:</u> The characteristics of materials determined with the polarizing microscope are:

(1) The birefringence or "amount" of double refraction.
(2) Thickness of thin sections.
(3) Extinction angle or direction of a doubly refracting material.
(4) Optical axis determinations.
(5) Right- and/or left-handedness of crystalline structures.
(6) Dichroism — pleochroism — an optical selective absorption characteristic dependent upon crystallization.
(7) Crystal system determination.
(8) Dispersion — characteristic for various colors of light and the crystal system.
(9) Refractive index.

Although most of the quantitative work and characterization by a polarized light microscope are performed on thin rock sections and minerals, numerous other materials may be characterized as well. They include crystalline aggregates, starch grains, natural fibers, films, foils, textile fibers, emulsions, suspensions, grains and spherulitic structures, and inclusions.

3. <u>Characteristic indicators</u>: With the polarizing microscope (Figure 68), some of the characteristics are directly indicated. Others must be either inferred from the observation and/or calculated from measurements. The visual indicators in this type of microscopy are:

(1) Presence or absence of light.
(2) Colors of light.
(3) Patterns of light, either colored or monochromatic.

Two methods of examination are used with this type of instrument. One is observation with plane polarized light, wherein the light from the source, and thence from the polarizer, passes through the object in essentially parallel rays. Examination is made with and without the analyzer in this case. Presence or absence of light and colors are indicators with this method. The other method uses convergent light through the specimen object,

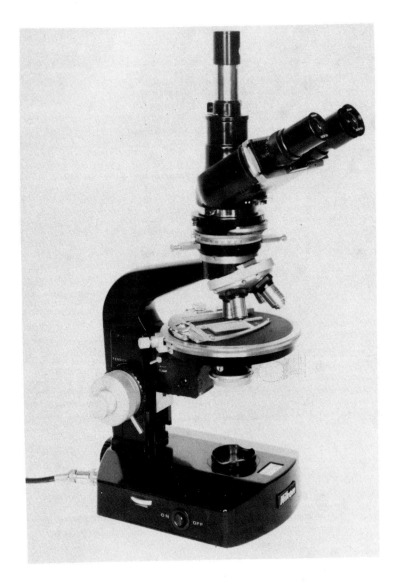

Figure 68. Research polarizing microscope. (photo courtesy Nikon.)

POLARIZED LIGHT MICROSCOPY 195

accomplished by using condenser lenses beneath the stage of the instrument. Viewing the pattern produced by the convergent light is termed conoscopic observation and is accomplished by examining the back focal plane of the objective with the regular ocular and a modifying optical element called the Bertrand lens (Figure 69). The latter changes the optics of the ocular such that instead of focusing on the real image from the objective, it focuses on the back-focal plane of the objective. This method of examination makes use of characteristic patterns produced by convergent polarized light in certain substances.

The actual quantitative characterization of substances examined by the polarizing microscope is accomplished by accurate measurement of angles (using the rotating stage and cross hair equipped ocular), distances of pattern elements in the image (usually with reference to the center of the field), and accurate assessment of colors produced in the images observed. The mere presence or absence of light is in itself, under certain circumstances, a definite characterization. For measurements involving thickness of thin sections or refractive index, the color quality (shades, hue etc.) is of prime importance. The colors are compared with standard charts and analysis made on that basis. Because the thickness of thin sections from one to another varies, wavelengthwise, the colors produced by the interference of white light components vary in "orders". That is, there is a periodic spectrum produced such that the same or nearly the same colors will prevail under several different thickness conditions. It is important in quantitative work to not only be able to determine the color (compared with standard color charts), but to determine the "order" of the spectrum within which it lies, which is dependent upon the thickness. An accessory called a quartz wedge (Figure 70) is commonly used in that determination. This furnishes a prismatic section of varying thickness of known orientation. A sequence of colors occurs in the wedge as the thickness increases. The order is generally violet, blue, green, yellow, orange and red. The sequence is repeated distinctly three times, and then as the thicker end of the wedge is approached the colors become fainter and not so clearly defined. The colors are divided into orders as mentioned above. The thickness of the wedge is marked in millimeters at intervals. Using the wedge, and observing and comparing colors, the thickness of a section, its refractive index and orientation may be determined.

It is beyond the scope of this book to treat the complex field involving the many uses of the polarizing microscope. Two other

CONVERGENT
MONOCHROMATIC LIGHT
OF SODIUM

Back focal plane of objective is
examined with aid of Bertrand lens.

Black pattern indicates
an assemblage of all points
where there is a difference in
phase of multiples of a wavelength.

CHARACTERIZATION BY CONOSCOPIC
POLARIZED LIGHT

Figure 69. Biaxial interference figure (axial plane parallel to vibration direction of polarizer).

volumes of this series are devoted exclusively to polarization phenomena using the microscope.

INTERFERENCE MICROSCOPY

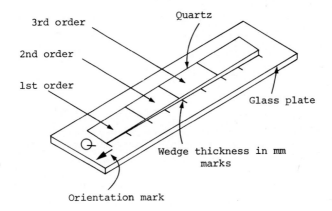

Figure 70. Quartz wedge.

D. <u>INTERFERENCE MICROSCOPY</u>: For purposes of this discussion, the basic operating principles of various types of interference microscopes presented in a previous chapter are adequate. At that point we discussed this form of microscopy from the standpoint of improved contrast. Interference microscopy, in its various forms, has greatly advanced the observational capability of the microscopist through increased contrast and the ability to discern detail otherwise invisible. Interference phenomena themselves, either used as a tool or as effects produced by subject material, also may characterize microscopic subjects. It is this aspect that will be treated briefly here.

1. <u>Characteristics determined</u>: The characteristics of materials examined by interference microscopy methods are:

    a. <u>Measurements</u>

        (1) Dimensions (length, width, thickness).
        (2) Dry weight and mass of cells in biological applications.
        (3) Rates of growth, organic and inorganic.
        (4) Surface finish and degree of wear in engineering applications (smoothness, roughness, flatness etc.)
        (5) Refractive index.

    b. <u>Morphology</u>

(1) Surface features.
  (2) Internal structure.

c. Structural analysis

  (1) Detection of weak structural points.
  (2) Stress concentrations.
  (3) Inclusions — voids and cracks.
  (4) Fracture examination.

2. Characteristic indicators: There are several different visual indicators that characterize subject materials when examined using interference techniques. The visual effects are dependent upon the particular type of interference technique used.

Interference instruments based on the Jamin-Lebedeff system are fundamentally measuring instruments and the visual indicators are in the form of interference fringes. For measurement purposes the monochromatic light of sodium or mercury is generally used. The interference patterns, their disposition and distance from one another are measureable. Specifically such measurements of interference patterns are used in, but not limited to, the study of the microtopography of crystals, semiconductors, ceramics, electroplated and vacuum deposited layers, photographic emulsions and magnetic recording tapes. The spacing of the interference fringes is indicative of the surface smoothness, for instance. Regular equally spaced fringes indicate a smooth unvarying surface. Irregularity in the fringes, in either their form or spacing, indicates variations in surface topography such as "hills and valleys" etc. (refer to Figure 71).

An example of measurement by interference is the determination of film thickness. In the case of opaque films, interpretation is straightforward. It is necessary to observe a boundary of the film so that the height of the base surface may be compared with the top surface of the film.

In the case of a transparent film on a glass or other transparent base, much more light will be reflected from the top of the film than from the back so the same method as that used for opaque films can be used.

With a metal film on a dielectric substrate, the method is also straightforward for films more than $0.5\ \mu$m thick, but with thinner films a correction must be made to allow for the difference in phase between light reflected from the metal and dielectric surfaces. In practice this correction can be determined once for a particular type of film by remeasuring after metallizing

INTERFERENCE MICROSCOPY

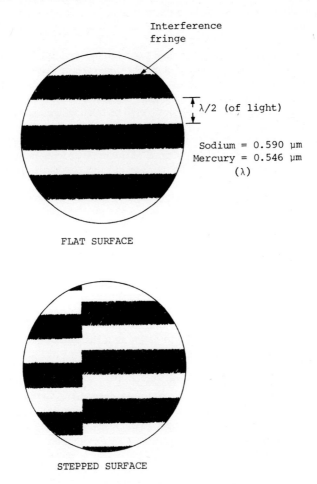

Figure 71. Characterization of surfaces by interference.

over the whole surface to eliminate the phase shift. The correction can then be added to the measured thickness of other films of the same kind.

With transparent film on a highly reflecting substrate, most of the light is reflected from the back of the film but the small amount reflected from the top surface introduces an error which is significant in films less than 1 μm thick and metallizing is necessary to obtain a direct measurement of the step height. The correction cannot be used for other films as it varies with film thickness.

If the film is more than two micrometers thick, separate white light fringe patterns can be observed on the top and back surfaces of the film (see Figure 72) so that the optical thickness can be measured without observing a boundary. The light from the top surface usually gives rise to faint fringes advanced by 2t, and that from the back surface gives bold fringes retarded by 2t (n-1) where "t" is the film thickness and "n" is its refractive index. To do this the positions of the central fringes are noted on an ocular reticle and the number of monochromatic fringes between these points counted.

The two-beam interferometry system, as exemplified by the Jamin-Lebedeff arrangement previously described, can be improved by the use of multiple-beam interferometric methods.

Multiple-beam interferometry makes possible the accurate measurement of surface topography to molecular dimensions. This is a system whereby a large number of beams from two surfaces can be combined to produce fringes. When this is done correctly the fringes can be sharpened up to narrow fine lines instead of the broad ones produced by the two-beam method. Instead of the fringe width being perhaps a tenth of the spacing between fringes, it can be as little as a fiftieth, thus improving the accuracy of measurement.

The multiple-beam technique can only be applied, however, to surfaces which have inherently a natural smooth polish. A number of specimen materials are suitable under these conditions. Among them are crystals, polished metals, glasses, plastics, smooth fibers and other polished surface materials. Matte surfaces do not respond to this technique.

The multiple beams are secured by coating surfaces with highly reflecting films of silver by vacuum deposition, from solution or electrolytically. Illustrated by Figure 73, the essential scheme is revealed. The slide and coverslip used are provided with partially reflecting coatings, and a pinhole is required at the first focal plane of the condenser. If a 25 mm objective is used as a condenser, a pinhole of 0.5 mm at its focal plane will be adequate. Light is partially transmitted and partially reflected at each of the metallized surfaces. In areas where the optical path difference between successive transmitted rays is an integral number of wavelengths, constructive interference obtains and a bright fringe is seen. A single fringe, therefore, traces out a

INTERFERENCE MICROSCOPY

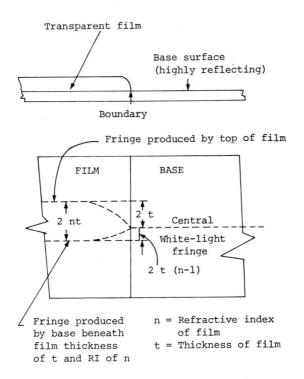

Figure 72. Transparent film thickness characteristics. (drawing adapted courtesy M.E.L. Equipment Company Ltd., Watson, The Microscopy Div.)

contour line of equal optical path and the deviation of a fringe as it passes through a specimen can be used to measure the optical path of the specimen.

Even by the above brief description of some of the matters involved in the use of interference microscopy as a measurement, morphological and structural analysis method, it is evidenced as being quite complex. Considerable training and experience in such techniques are prerequisite to expertise in the field. Other volumes of this series furnish the necessary detailed background.

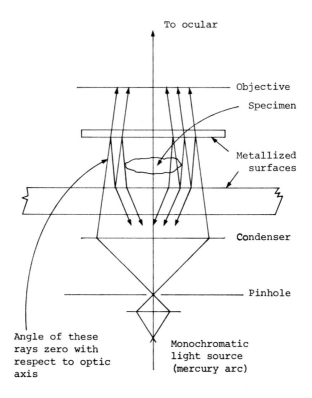

Figure 73. Essentials of multiple-beam interference setup.

E. FLUORESCENCE MICROSCOPY

1. The fluorescence microscope: Characterization of materials by light excitation results in the fluorescence phenomena previously defined.

Referring to Figure 74, the essential elements for producing and observing fluorescence phenomena with a microscope are shown. The elements peculiar to such a microscope are a special high intensity lamp whose output is rich in the ultraviolet portion of the spectrum, an exciter filter, a darkfield condenser and a barrier filter interposed between the objective and ocular.

FLUORESCENCE MICROSCOPY

The exciter filter allows a selected portion of the emitted spectrum from the lamp to be applied to the specimen. The specimen responds to the stimulating radiation and fluoresces in accordance with its characteristics. The barrier filter excludes transmission of radiation from the exciting source (the lamp) and passes only the fluorescing radiation from the specimen.

In most cases, normal optics of glass construction are used in such microscopes. However, certain applications wherein very short wavelengths of light are employed which would normally be absorbed by glass, then quartz optics, first-surface illumination mirrors, quartz microslides etc. may become necessary.

Because of the low percentage of incident light converted into fluorescent light, the exciting lamps are normally of very high intensity. A high pressure 200 watt mercury-arc source is typical. However, ultraviolet light is not always necessary as any light radiation shorter in wavelength than that of the anticipated fluorescence may be used, even in the visible range.

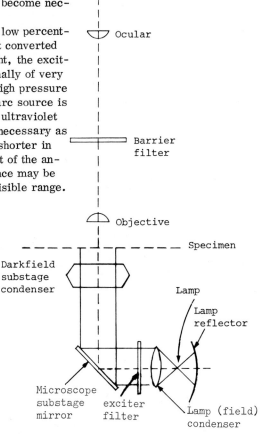

Figure 74. Basic fluorescence microscope.

2. Characteristics determined: Some of the characteristics determined by the use of the fluorescence microscope are:

  (1) Chemical composition.
  (2) Affinity for certain stains.
  (3) Quantitative and qualitative response to stains.
  (4) Transport and distribution of metabolites in histological studies.
  (5) Cell differentiation.
  (6) Fiber structure.

3. Characteristic indicators:

  (1) Presence or absence of fluorescence (bright against dark background)
  (2) Affinity for specific stains by locality within or on structures.
  (3) Movement of specific fluorescing elements.
  (4) Special distribution of the fluorescing phenomena, both spatially and spectrally.

Two basic divisions may be made in the exhibition of fluorescence microscopically, primary and secondary.

Primary fluorescence is mainly a functional characteristic of the chemical constitution of a substance. It either fluoresces (at a characteristic wavelength) or it does not in the presence of exciting radiation. A well known example of primary fluorescence is that of chlorophyll which fluoresces a brilliant red. Proper diagnosis of these phenomena is useful in making chemical composition determinations of specific sample materials. In making these determinations of the various fluorescent substances, use is made of the spectral distribution of the fluorescent phenomena. A pupillary spectroscope may be used in this respect, permitting selection of various barrier filters adaptable to the spectral emission of the fluorescing substance. These methods can be and are easily applied to materials that are structureless — liquids for example. The advantage of making determinations in this way is in the extremely small sample required.

The secondary methods of fluorescence make use of special fluorescent stains termed "fluorochromes". Fluorochromes brightly fluoresce when exposed to proper exciting radiation.

Various tissues and substances show characteristic or specific affinity for certain fluorochromes. Different parts of a

FLUORESCENCE MICROSCOPY

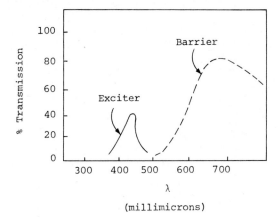

Figure 75. Spectral transmission of typical exciter and barrier filters.

specimen may show a different qualitative and/or quantitative reaction to a fluorochrome, just as common stains exhibit differentiation. One of the most outstanding of these vital stains is acridine orange. Fluorescence created by some such dyes (fluorochromes) which form pure chemical bonds (not adsorption complexes) with the antigen (infectious agent) protein substances, which are collected by certain animal or plant organs, are easily detected with the fluorescence microscope (Figure 76). A most frequently used fluorescing dye for this purpose is fluoresceinisothiocyanate (FTIC).

Applications of fluorochromes to microscopical examination are especially valuable in histology in tracing the transport and distribution of various metabolites (absorption, imbibition, secretion, excretion, transportation etc.). It is possible in many cases to fluorochrome organs in full function, without damage or impairment, providing a powerful tool for examining the dynamics of living organisms. In hematology such application can aid in counting and differentiating leucocytes, in bacteriology to provide streak differentiation contours, and in parasititology to demonstrate flagellata, sporozoa and virus types.

An example of an advanced technique is that of cytofluorometry. This specialized procedure is used in the quantitative determination of nucleic acids, proteins, amino acids, biogenic

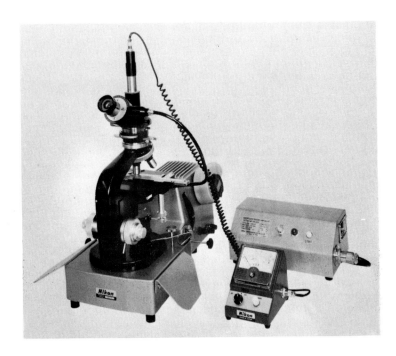

Figure 76. Fluorescence microscope with high intensity illuminator and measurement facilities. (photo courtesy Nikon.)

amines as well as substances added to cells and cell organelles for test purposes.

The quantity of the substance is determined simply by measuring the light emitted by all parts of an object in the exit pupil of a microscope. In contrast to absorption photometry, fluorometry at first supplies only relative values for substances. Absolute values are obtained by comparison measurements with known substance quantity.

Essentials of an arrangement for making such measurements are illustrated in Figure 77. The specimen is observed in phase contrast and measured by swinging out the observation mirror from the optic axis of the instrument. A complete description and

FLUORESCENCE MICROSCOPY

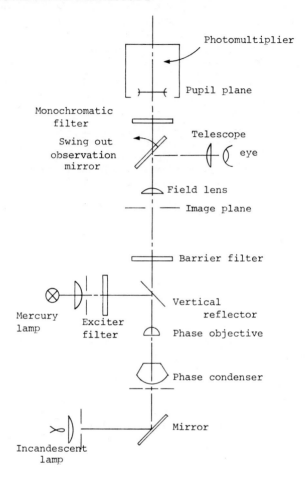

Figure 77. Essentials of Zeiss cytofluorometer. (drawing adapted courtesy Carl Zeiss, Inc.)

application of this particular instrumentation is referenced at the end of this chapter.

The phase contrast technique is also combined with fluorescence microscopy to produce a superimposed phase contrast image with that of the fluorescent image simultaneously to the observer. To obtain this effect, a lamp illuminator equipped with a built-in beam splitter, mixing light consisting of exciting UV and an adjustable portion of tungsten light, is used.

Appropriate components of both the phase contrast and fluorescing microscope can be incorporated into the same stand. E. Leitz currently markets such a combination.

F. MICROHARDNESS TESTING: Another characteristic of materials is that of hardness. The concept of hardness has been extended to include extremely small optical dimensions and, although the complex "hardness number" represents a highly diverse set of material properties, it is used extensively and particularly in the metallurgical field.

Microhardness is generally defined as the ratio of the load to the contact area of the impression. A diamond pyramid, or cone, is commonly used, being pressed against the specimen with a uniformly increasing load. The ratio of the load to the area of the permanent test impression is calculated. The measurement is always made after removal of the load. The microhardness range is considered to encompass a hardness number (P) wherein its value is equal to or greater than zero and less than or equal to a value between 50 p to 200 p where p is 1 pond or 1 gram force (gf).

Since the inhomogeneities of a given material are comparable in dimension to the diagonal dimension of the impression in the microhardness range, microhardness numbers are sometimes not very reliable quantities and not simply related to hardness values obtained at larger scales. However, they are of considerable value wherein they are relatively compared. Figure 78 shows a series of microhardness indentations as a measure of case hardening in metal.

Two common hardness designations are the Vickers and the Knoop. The indenter for the Vickers hardness is in the form of a square-based pyramid and that for the Knoop a rhombic pyramid. The depth of the impression (h_r) made by the square-based pyramid is related to the diagonal distance (d) as 1/7 times d, and for the Knoop indenter as 1/30.5 times d_{max} where d_{max} is the long diagonal.

Appropriate equations and tables are used, after diagonal measurement, to obtain microhardness numbers. As all materials are elastic to some degree the diagonal measured, after removal of the load, is incorrect by the amount of elastic recovery. When exacting figures are desirable, the latter factor must be taken into consideration during the calculations.

In making measurements of microhardness, because of the

MICROHARDNESS TESTING

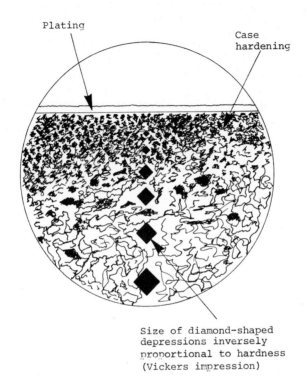

Figure 78. 150X hardness profile using microhardness tester.

anomalies present at such small dimensions, a visual qualitative evaluation of the indentation and a timed application of load are recommended procedures.

Essential accessories for such measurements are:

(1) An ocular reticle, or preferably a screw micrometer ocular, for measuring the diagonal(s).

(2) A microhardness tester comprised of an indenter and load measuring device or indicator.

Zeiss markets a very complete set of accessory parts for microhardness testing applicable to their Universal-type stands.

Their microhardness tester incorporates a special purpose vertical illuminator and electrically controlled shift mechanism for the objective and diamond indenter. The objective furnished is a 63X, NA 0.9 lens mounted side-by-side with the indenter in the tester assembly. The indenter supplied is of the Vickers type, but the apparatus may be adapted for the Knoop rhombic pyramid as well.

The loading speed is checked by watching a needle deflection on a galvanometer display unit. During measurement the entire system carrying the objective mount is moved back and forth between diamond tip and objective axis by means of an electric motor drive. The reproducibility of the setting is claimed to be better than plus or minus 0.3μm. For accurate calibration of the measuring ocular, an accessory drawtube-type monocular tube and a stage micrometer are used.

For greatest contrast enhancement in obtaining hardness characteristics, differential interference contrast (DIC) of the Nomarski type is often used. This method provides for a variety of contrasts in various colors to improve accuracy in the measurement of the diagonals, photomicrographic recording and for qualitative assessment of the specimen and indentation. Ordinary brightfield, darkfield and polarizing light illumination methods are also used in such measurements.

The major reason for utilizing the most advanced techniques available for obtaining contrast in this type of characterization of a material is the requirement for specimen and indentation evaluation. Despite the apparent simplicity of the method, the results involve a rather complex set of conditions. As mentioned before, the scale of the indentation is comparable, in many cases, with the inhomogeneities of the material. Thus, the impressing surface of the indenter may encounter a number of different individual components, all reacting differently to the load. Also, even if the material is entirely homogeneous, there are two phases which obtain under action of the indenter. One is elastic and the other plastic. While the elastic deformation component represents a reversible change in shape, the material is deformed permanently by slip, twinning etc. in such a way that the mass deficit of the indentation piles up at the edges. Because of these factors the indentation resulting is usually asymmetrical in geometry, and even its surfaces may not be plane ones. For these reasons, it is left to the observer to estimate the ridge surrounding the indentation and determine just where to delimit the length of the diagonal to be measured. The evaluation of the

MICROHARDNESS TESTING

ridge, therefore, plays an important part in the accuracy of the length measurement and the utmost in visibility conditions becomes paramount.

Differential interference contrast is superior to any other method for enhancing contrast under these conditions as minute level differences, as may be produced by different hardness, are made visible in the form of relief, either in black and white or color contrast.

There are a number of ways in which the microhardness of a material is evaluated. A complete discussion of them is beyond the scope of this brief treatment. Perhaps the most common method is by computing a microhardness number from:

$$HV \text{ (kp/mm}^2\text{)} = 1854.4 P/d^2$$

or: $HK \text{ (kp/mm}^2\text{)} = 14230 P/l^2$

where HV = Vickers hardness (square based pyramid)
HK = Knoop hardness (rhombic pyramid)
P = load in grams
d = mean diagonal length in μm
l = length of long diagonal in μm

In addition to microhardness characteristic determinations, the indenter can also be used to provide azimuth information in regard to crystal surfaces. If crystals are indented on particular surfaces, the directions of axes and cleavage planes are revealed relative to the indentation diagonals. Angular measurements of these relationships are characteristic of particular materials.

In all of the applications of microhardness accessories, the accurate leveling, or rather the orientation, of the surface under examination or test is of primary importance. For the elimination of measuring errors it is essential that there is an accurate adjustment of the surface under examination so that it is normal to the optic axis of the microscope.

G. <u>CHARACTERIZATION BY FORM AND ORIENTATION</u>: Materials and/or specimens are often characterized by their form. Form is defined as the shape and structure of anything, and is perhaps one of the different modes or aspects of existence or manifestation of the same thing or substance. For instance, graphite and soot are forms of carbon. In general, form is the aspect under which a thing appears as distinguished from substance or color. In this discussion we will be concerned mainly with the external appearance, shape, form factor, aspect ratios and morphological features in general.

1. <u>Chemical means</u>: A great many substances may be characterized by form using chemical methods to change their mode of existence into recognizable and typical shapes. This type of characterization is used extensively in the study of mineral and other chemical compounds which adopt certain crystalline forms and habits due to chemical reaction.

Chemical microscopy is concerned to a large degree, although by no means exclusively, with crystalline manifestations of chemical reactions under the microscope. Figure 79 illustrates some microchemical reaction results in typical crystal

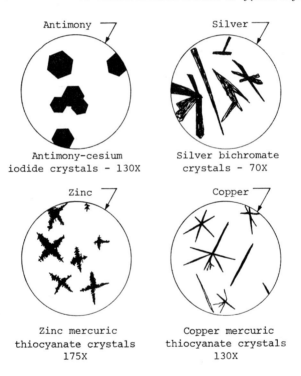

Figure 79. Chemical characterization by form (crystal habit). (adapted from microscopic determination of Ore Minerals Geological Survey Bulletin 914.)

habits. This type of characterization is used extensively in the identification of minerals, qualitative analyses of very small specimen quantities and in the investigation of poisons and narcotics. The methods in chemical microscopy incorporate hot and cold stages (as accessories), and considerable other specialized apparatus for microchemistry far too numerous for inclusion here. Volume 43 of this series is devoted exclusively to the subject.

2. Crystal rolling: The habit of a crystal is defined as the characteristic shape, as determined by the crystal faces developed, their shapes and relative proportions.

Therefore, for accurate determinations of the habit, a crystal should be examined in all of its aspects. That is to say, it should be looked at from as many different angles as necessary to assemble all information in regard to its features. In microscopy, the crystals that are commonly examined are very small and in many cases do not lend themselves, readily at least, to manipulation and detailed examination from every side. A technique that is useful for the purpose in many cases is that of crystal rolling.

If a temporary water-mount of specimen material is placed under the microscope, it will be found that the coverslip in most cases can be "slid" with the point of a needle to one side on top of the fluid, uncovering a portion of the specimen material without disturbing its position or location in the field. In fact, this phenomenon is often used to select and remove certain portions of specimen material from the main body. This ability is largely due to the low viscosity of the water and relatively high inertia and/or specific gravity of the enclosed specimen elements.

The "reverse" of this condition can, of course, be made to exist. That is, if the viscosity of the fluid (instead of water) used is high, then slight movements of the coverslip with a needle will result in movement of the specimen with the fluid. The advantage of this mechanical fluid coupling condition, wherein high viscosity immersion fluids are used, is that crystals can be moved within the fluid by moving the coverslip in an appropriate direction. If the fluid is selected properly (viscosity etc.) in relation to the crystal material (size, specific gravity etc.), it will exert the proper forces to "roll" or turn the crystal along one of its axes to allow examination of most, if not all, surfaces.

The same attribute of high viscosity fluid can be used to examine microfossils in various aspects. Foraminifera and radiolaria are especially suited to examination in this way.

High viscosity immersion oils for the purpose are available commercially. R. P. Cargille Laboratories, Inc. markets such oils and provides specifications on refractive index, dispersion, temperature coefficient, stability, fluorescence (long-wave and short-wave ultraviolet), color, viscosity, density, cloud point, flash point and neutralization equivalent. Some of these properties are very valuable to know in the many applications of immersion fluids. Various index of refraction liquids likewise may be similarly employed. Proper selection of index of refraction as well as the viscosity is of prime importance for maximizing visibility of specimen details.

3. Rotation apparatus: The Hartshorne Rotation Apparatus (Figure 80) is designed to facilitate the determination of optical and morphological properties of crystals or liquids. It enables quick centering on the optical axis of the microscope and on the axis of rotation of the crystal.

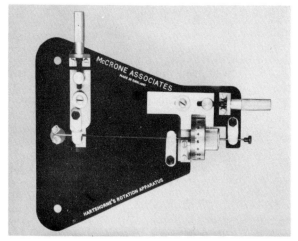

Figure 80. Hartshorne Rotation Apparatus. (photo courtesy Walter C. McCrone Associates, Inc.)

The apparatus functions best with crystals that are not too small to be manipulated individually (approximately 0.2 mm in diameter or larger). This "spindle stage", as such a device is sometimes called, allows a crystal, after being suitably attached

CHARACTERIZATION BY FORM AND ORIENTATION   215

with an adhesive, to be rotated within a liquid of known refractive index. Also, if mounted on a rotating microscope stage, another axis of rotation is then provided.

Observation of the various phenomena obtainable with the device is, of course, dependent on the light characteristics applied (nonpolarizing, polarized etc.). Both conoscopic and/or orthoscopic observation are of use with the apparatus. Its construction allows angles to be easily determined to the nearest degree from a drum mounted concentrically with the axis of rotation.

By conoscopic observation of typical interference figures or by stereographic plots of extinction angles, the refractive index of crystals at various orientations can be determined.

The liquid in which the crystal is rotated is contained in a small cell integral to the rotation apparatus, and is held there by capillary action.

With some knowledge of crystallography an investigator, using this apparatus, can determine in addition to refractive indices various crystallographic axes. Also, methods have been devised using this apparatus to determine the unknown refractive indices of liquids (which would be contained in the cell) by observing the action of known crystal rotation.

4. *Universal stage*: The ultimate device for precise orientation of specimen material under the microscope objective is perhaps the universal stage. This complex device is usually furnished in a four-axis rotary version with centration and specimen vertical adjustments. Accurate scales provide means for measuring angles of tilt and rotation in practically any aspect. Conoscopic observation is commonly used with this accessory, and measurements made are plotted on a sterographic network for record and analysis purposes. Because of the size of such a stage in the elevation dimension, special long working distance objectives with short mounts are normally used. As the specimen slide is normally mounted between two glass hemispheres (one above and one below), the objective's magnification and numerical aperture apply when they are used with hemispheres of the proper refractive index. Zeiss, with their universal stages, furnish large and small hemispheres with refractive indices of 1.557 and 1.648. The arrangement using the hemispheres enables light to pass through the combination without any disturbing refractions.

In mineralogy, the universal stage is used to determine the x, y and z directions, the optical sign and the axial angle. The x, y and z directions are found by orientation of the stage that shows uniform extinction while the section is being rotated on a horizontal axis. After they have been found, the position of the axial plane is established. Then, by rotating the section on the y direction as an axis, the directions of the optical axes can be made vertical. When an optic axis is vertical the section will remain uniformly extinguished during rotation around a vertical axis. The optical sign and the axial angle can then be determined.

Details of the use of this accessory is beyond the scope of this book. Indicative of its complexity, an entire volume of this series is devoted to the Universal Stage.

H. DISPERSION STAINING: This is a microscopical technique based on the dispersion properties of various substances, and has become a very powerful diagnostic tool in the hands of the experienced microscopist. It is extensively used in all fields requiring small particle identification, particularly in the industrial health and hygiene area.

The dispersion staining effect arises only when there is an interface between two transparent phases having different dispersion curves, but identical refractive indices somewhere in the visible spectrum (Figure 81).

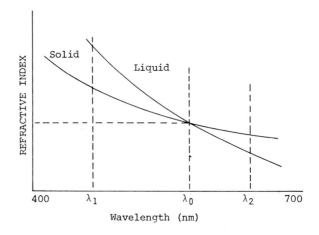

Figure 81. Dispersion curves. (illustration courtesy Walter C. McCrone Associates, Inc.)

DISPERSION STAINING

1. <u>Darkfield method</u>: An ordinary laboratory microscope can be used in this method. It is essentially a technique of darkfield microscopy, and only one accessory, a darkfield stop located in the lower focal plane of the condenser, is needed to demonstrate the effect. The stop diameter should be of the proper size to provide a darkfield (refer to Chapter 2), and the condenser should be corrected for chromatic aberrations.

   a. <u>Operating principle</u>: A microscope equipped as described above produces a darkfield view at the ocular. A cone of light is focused at the specimen site by the condenser at such an angle that only rays deflected by interfacing substances having different refractive indices will enter the objective. A specimen having a refractive index the same as the medium it is mounted in will be invisible as there is no visual evidence of the specimen. If the specimen is of a different refractive index than the mounting medium it will be visible, in a darkfield, at the ocular.

   If white light is used at the source, composed essentially of red, yellow and blue components, proper selection of mounting media (immersion fluids) in regard to the specimen material will produce dispersion staining effects.

   As the refractive index of a substance is different for differing wavelengths (colors) of light, the selection of media is made on that basis. For instance, referring to Figures 82 and 83, if a quartz specimen particle is immersed in a fluid for which the refractive index is 1.544 for yellow light, that wavelength of light will not encounter a discontinuity upon entering the quartz as its refractive index is also 1.544. If, for the remaining components of the incident white light (red and blue), the quartz offers a different refractive index than the medium, the quartz will appear in those colors (red and blue) in relative degrees dependent upon its crystallographic axes and their orientation with the optical axis of the microscope. The image is then termed "dispersion stained", as the effect is dependent upon the relative dispersion of component light waves in the immersion medium and the subject material. Since the method is based on the use of liquids of much greater dispersion than the specimen material, the greater the difference, the more brilliant the colors are rendered in the image at the ocular.

   The yellow light portion of the spectrum passes through the optical axis of the microscope at an angle such as to not enter the objective. The red and blue are refracted at the quartz-liquid interface as to enter the objective. Since the blue end of the spec-

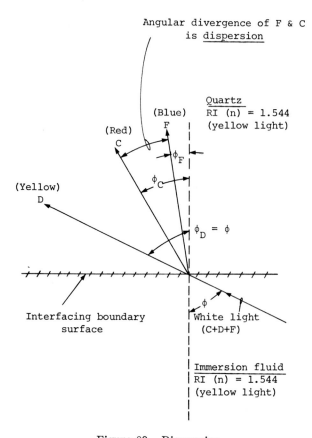

Figure 82. Dispersion.

trum is refracted to a greater degree (more toward the objective axis) than the red, the quartz particle in this case will appear largely blue with a slight tinge of red. If the red is difficult to distinguish, its visibility can be enhanced by slightly decentering the darkfield stop.

In this method a cap analyzer can be used in the identification of anisotropic substances. Isotropic (amorphous) substances and crystals of the isometric system can have but a single value of refractive index. Crystals of the tetragonal and hexagonal systems have two principal refractive indices corresponding to the velocities of light having definite directions of vibrations. Fur-

DISPERSION STAINING

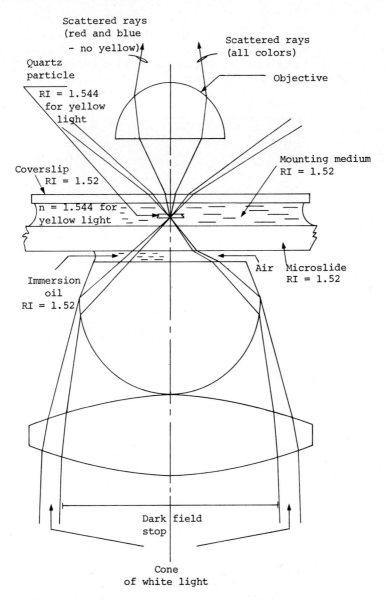

Figure 83. Dispersion staining darkfield method.

ther, all orthorhombic, monoclinic and triclinic crystals have, similarly, three principal indices.

2. Dispersion staining objective: Another approach to dispersion staining does not use the darkfield microscopy technique described previously, but a specially constructed objective. The objective is pictured in Figure 84 and a schematic illustrating its operation in Figure 85.

Figure 84. Dispersion staining objective. (photo courtesy Walter C. McCrones Associates, Inc.)

White light is supplied as before, but instead of being interrupted by a central stop in the substage lower focal plane, it is adjusted to be essentially axial by closing down the substage condenser iris to a very great degree. This axial pencil of white light then proceeds through the object plane where the specimen material is immersed in the dispersion liquid and thence through the objective. In the specimen plane the specimen interface with the immersion fluid acts in the same manner as before. If it is quartz immersed in a liquid with a refractive index of 1.544, the yellow portion of the incident white light will pass through the interface unrefracted and the blue and red portions will be refracted. (It is understood that the interface is at some angle other than normal to the direction of the incident light.) At the back focal plane of the objective, the components of the incident white light so dispersed will be spaced proportionate to the degree of

DISPERSION STAINING

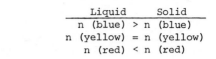

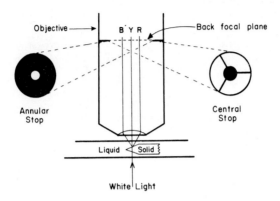

Figure 85. Light path through dispersion staining objective. (drawing courtesy Walter C. McCrone Associates, Inc.)

dispersion. At this point a diaphram with an annular stop, or a central stop, is placed. The annular stop allows only the central or yellow rays to pass and the central stop conversely allows the red and blue rays to pass, blocking the undeviated yellow. Thus, at the intermediate image plane the object is "dispersion stained". The dispersion staining objective is designed so that either, or neither, of the stops is in place, by rotation of a turret holding the stops in the objective back focal plane. The objective is a 10X 16 mm, 0.25 NA lens system. A detenting action provides for positive selection of the various modes of operation — annular stop, central stop or full normal aperture.

3. <u>Characteristics determined</u>: Dispersion staining is used chiefly in particle identification through the determination of refractive index, separation of amorphous and crystalline substances, and identification of crystal classes.

4. <u>Characteristic indicators</u>: Changes in colors, degree and quality of "dispersion staining" are the direct indicators of mate-

rial characteristics using this method. If the solid specimen particle and the liquid in which it is immersed have refractive indices very different, dispersion colors will not be indicated in the image. Instead, the image of the specimen with the central stop will appear white on the dark background. With the dispersion staining objective, it will show black edges on a brightfield with the annular stop and white edges in a darkfield with the central stop.

Table X lists the colors observed using either the annular or central stop with the dispersion staining objective. The colors

TABLE X

Dispersion staining colors

Matching λ_0 nm	Annular stop colors	Central stop colors
<420	blue-black	light yellow
430	blue-violet	yellow
455	blue	golden yellow
485	blue-green	golden-magenta
520	green	red-magenta
560	yellow-green	magenta
595	yellow	blue-magenta
625	orange	blue
660	orange-red	blue-green
>680	brown-red	pale blue

shown are a continuous series throughout the entire range of λ_0 (Figure 81). This serves as the basis of a method of determining λ_0 by comparing the colors produced when the annular and central stops are used with those listed in Table X.

5. *Special considerations:* Problems of obtaining brilliant colors with certain specimen materials or with very small particles are mainly concerned with the compounding of special liquids with very high dispersions even in the lower refractive index ranges (below 1.55). Considerable research along these lines has produced a long list of applicable fluids.

Increased visibility of very small particles through more brilliant coloration can sometimes be effected by a heating stage as many liquids change their refractive index at easily detectable rates with a change of temperature. The RI of the liquid decreases with increasing temperature. As the preparation is heated colors are obtained ranging from red, orange and yellow through green to blue (annular stop).

The dispersion staining technique is comparatively easy of accomplishment, but the interpretation of results, in which coloration plays a most important part, is for experts and considerable experience is required for accurate results.

Nevertheless, the addition of identification, even in the rather rough classifications of amorphous, isotropic and anisotropic substances and determinations of RI for even fractions of particle populations, provides enormous advantages over ordinary size distribution studies. It is a very versatile technique and can be used for many problems of identification of microscopical quantities of matter. Volume 35 of this series is devoted entirely to the subject.

I. ULTRAVIOLET MICROSCOPY: In Chapter 1 a microscope equipped to use ultraviolet light as the specimen illuminant to increase resolution of microscopic structure was described. At the turn of the century forerunners of that type of instrument, with optics developed first by Köhler and Von Rohr around 1900 and several systems made by Zeiss in 1904, were mainly directed toward that goal.

Since the early 1930's, however, especially with the advent of the electron microscope, the emphasis on ultraviolet microscopy is on specimen characterization rather than on increased resolution. It was found that various biochemicals demonstrate marked absorptions at various wavelengths throughout the ultraviolet region. For example, 10% nucleic acid (purine and pyrimidine) in a thickness of 5 μm has an absorption of 90% at 260 mμ, while 10% protein under the same circumstances has an absorption of only 2%.

Characterization of many different types of biological materials (even live ones) such as stained sections, tissue cultures and individual cells are possible by measurement of UV absorption. The balance of certain biochemicals in such material may, for instance, be changed by environmental conditions, exposure to certain other radiation or by abnormal growth conditions. Measurement of these changes through variations in UV radiation can characterize both normal and abnormal conditions of the material.

In addition, there is now a wide industrial application of the ultraviolet microscope to characterization of polymers and fibers.

The same approach is used in this field as in biology. The absorption characteristics of polymer systems, whether in granules, fibers, films or molded samples, are measurable and thereby important attributes. Examination of materials before, during and after certain treatments are specific in their reaction absorptionwise to UV. This type of research leads to improved processing, better final materials and improved understanding of fundamental physical and chemical relationships.

J. MICROPHOTOMETRY

1. Introduction: Photometry is concerned with the determination of the transmissivity (or the degree of extinction) of objects viewed in transmitted light, the determination of the reflectivity in reflected light, fluorescence intensities by both transmission and reflection microscopy, and the determination of path differences in interference microscopy. The results determined constitute characterizations of materials and/or specimens (see Figure 86).

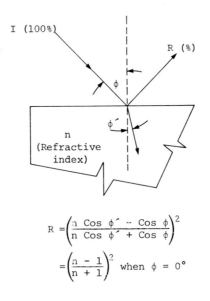

$$R = \left(\frac{n \cos \phi' - \cos \phi}{n \cos \phi' + \cos \phi}\right)^2$$

$$= \left(\frac{n-1}{n+1}\right)^2 \text{ when } \phi = 0°$$

Figure 86. Incident (I) and reflected (R) light relationships.

MICROPHOTOMETRY

There are numerous applications of microphotometry. Among them are:
- (1) Reflectance of mineralogical and metallic surfaces.
- (2) Transmission of thin films.
- (3) Optical densities of photographic films.
- (4) Localized and/or integrated density of cells and tissues.
- (5) Fluorescence intensities.
- (6) Morphological presentation.
- (7) Intensity variations based on spectral differences.
- (8) Specific spectral absorption of biological and industrial products.

2. Instrumentation: Figure 87 illustrates the essential modern elements for using a microscope for photometric purposes in transmitted light. The term "modern" is used here advisedly, as many of the older methods are in use in narrow or restricted fields of investigation. For instance, in reflectivity measurements two older methods are of some use. One is by comparison, wherein a reference surface is compared with the one under examination and judged by eye, great reliance being placed on the experience of the investigator. Although the eye can detect minute differences in light intensity when the color of the two adjacent fields is the same, it is not so when the fields are of different colors. The other is the photoelectric method using a photoelectric cell positioned at the eyepoint of the microscope ocular. The two methods are not really comparable, however, because of the difference of spectral sensitivity between the eye and photocell. The older photocells used in photometry such as alkali and selenium cells have largely given way to the photomultiplier/detector tube because of its greater sensitivity and flexibility.

The means illustrated in Figure 87 provide, in essence, a basis for most microphotometric work with the light microscope: a high intensity light source, a light modulator, a selectable field stop, a photoelectric detector and an indicator (either analog or digital).

Because of the nature of photoelectric measurements, the wide range of densities in transmission work and the great variation in absorption and reflectance of materials investigated by reflected light, a high intensity light source is required.

The light modulator is usually of the "chopper" variety,

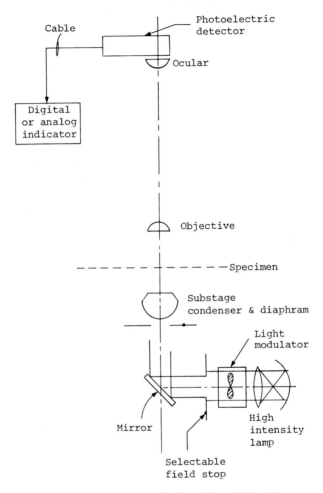

Figure 87. Essentials for microphotometry (transmitted light).

wherein the light is interrupted at a sufficiently high rate as to be applicable to electronic amplifiers. Direct current (DC) amplifiers, of course, could be used, but the major advantage of the "chopper" or interrupted light, and the use of more conventional AC amplification is that the diagnostic light is thereby isolated from any steady-state ambient light that might interfere with the measurement process.

The photoelectric detector commonly employed is of the photomultiplier type. This type of detector possesses a photoemissive cathode, which upon being exposed to light emits electrons, which are in turn attracted to a higher potential electrode. Each of these electrons upon striking that surface causes secondary emission of others which are then attracted to an electrode of still higher potential, and so on. This process, repeated several times, results in a highly sensitive detector whose current flow in the output circuitry can be quite large even for very low light intensity conditions at the cathode.

The selectable diaphram located at the source of illumination can be adjusted such that a very small selected area is illuminated in the specimen plane, providing greater repeatable accuracy on localized measurements. Either sets of fixed stops of great accuracy may be used, or some manufacturers supply variable stops that, depending upon the optical system, are continuously adjustable so that the measurable field is from a few millimeters to about $0.5\ \mu m$ in diameter.

The indicator, if an analog type, provides a meter indication of voltage or current that is proportional to the light intensity, and which may be calibrated and directly read in light intensity units. Digital indicators indicate directly those intensities with numerical displays.

Variations and expansions on the basic illustrated arrangement are available commercially. They range from merely providing a simultaneous viewing ocular, to fully automated and even fully programmable apparatus complete with printed output results.

In reflectivity measurements, instead of the light source being applied as illustrated, it is introduced into the microscope optical system via special coupling devices, such that it furnishes vertical illumination from above the specimen (which is opaque in most cases) in the same way that other reflected light illumination is accomplished. The modulated light is then reflected from the surface under examination and proceeds back up through the optical train to the photoelectric detector.

3. Characteristics determined

   a. Reflectivity: This quality is defined as the ratio of the intensity of light reflected from a given surface to the intensity of light vertically incident upon the surface. Minerals and metals are two of the chief materials rendering significant characteristic values of reflectivity. Some of them absorb very strongly at certain frequencies of light, resulting in colorations of mineral and metallic surfaces which are also important characteristically.

Leveling of the specimen reflecting surface for measurements can be very important, as recorded readings can vary as much as, or more than, 2% if the specimen surface is out of level as little as 0.004 inch.

Also, perfection of the polished surface has an important bearing on the measured reflectivity. Careful preparation of the polished specimen surface and considerable experience are required for the best results. If conditions are ideal, a reflectivity of, for instance, 37% can be attained with an accuracy of $\pm 1\%$ with modern microphotometers.

b. Transmissibility: Substances of all types (non-opaque, translucent or transparent) can be characterized to some extent by the degree of transmittance of light of monochromatic, mixed or selected bands of frequencies.

Light measurement in transmission through specimen material under the microscope is characteristically revealing in many cases.

Transmission measurements of the ultraviolet absorption of biological material and industrial products, for instance, is an important application of microphotometry. Contour maps of optical densities throughout a specimen can be used in internal diagnosis of it or for morphological purposes.

By the use of filters and monochromators, the measurement of light intensities can be extended to particular regions of the spectrum. Types of filters employed include, but are not specifically limited to, standard color filters for broadband illumination, and interference filters for spectral line and narrow band investigation. Prism or grating monochromators with variable slit widths, for width and range coverage control over the range of 230 to 700 nm, are also used.

c. Photometry versus stereometry: Photometric and stereometric methods overlap to some extent. However, they mostly supplement one another. It is apparent that variations in reflectivity or transmissibility can be revealing of surfaces (contours), shapes (outlines) and internal variations (densities etc.). Because stereometry is concerned with just those types of characteristics in the determination of geometrical parameters of objects (or object features) such as length, area, volume etc. photometric measurements can be directly used in those determinations.

Statistical parameters such as the number of objects or features can be determined in brightness distributions, size distributions, spatial distributions etc. Again, photometric measurements can be connected to these determinations.

However, for microphotometry to be very useful in stereometric determinations, it is necessary that a series of sequential photometric measurements take place in very large numbers. In that case, scanning devices wherein the scanning rates are high compared to those used in purely photometric determinations are necessary. In photometric scanning, the rate is relatively slow and the detection of tones (amplitudes) of light is accurate. In photometric applications to stereometric procedures, accuracies need not be as great and scanning rates can be considerably higher.

By means of so-called scanning photometers, wherein the coordinates of each point under measurement are determined, and using a suitable programmed computer, stereometric as well as photometric results can be obtained. Zeiss markets such a system.

d. Microspectrophotometry: Spectral measurements can be made microscopically with the use of a pupillary spectroscope. Light emerging from the microscope, whether transmitted through the specimen or reflected from it (in incident light microscopy), is examined spectroscopically. With the aid of this type of instrument, interference colors as the result of polarized light acting upon doubly refractive substances also can be examined. Positions of absorption bands and emission lines in the visible spectrum are data upon which direct or indirect inferences may be drawn as to specimen characterization. Microspectrophotometry is covered in detail in Volume 26 of this series.

K. REFERENCES AND COMMENTARY

1. Hausmann and Slack, Physics, Second Edition, D. Van Nostrand, 1939.

2. Condon and Odishaw, Handbook of Physics, McGraw-Hill, 1958.

3. Schaeffer, Microscopy for Chemists, Dover Publications Inc., 1966.

4. Ford, W. E., Dana's Textbook of Mineralogy, Fourth Edition, John Wiley and Sons, 1953.

5. Furman, N. H., Editor, Scott's Standard Methods of Chemical Analysis, Fifth Edition, D. Van Nostrand, 1939.

6. Microhardness Tester, Zeiss Catalog 41-700-e.
A brief discussion and description of microhardness testing and apparatus.

7. Jones, Francis T., "Refractive index determination for liquids by crystal rotation," The Microscope and Crystal Front 15, No. 8 (January-February 1967).
Immersion of a crystal that can be rotated in a fluid of unknown RI. The crystal of known properties is oriented properly and rotated to permit all values between its maximum and minimum refractive indices to be brought into known positions. Matching the known RI of particular crystal to the unknown liquid is done visually.

8. Hartshorne, N. H., "Single axis rotation apparatus and accessory devices for studying the optical properties of crystals," The Microscope and Crystal Front 14, No. 3 (November-December 1963).

9. Goodman, R. A., "Expanded uses and applications of dispersion staining," The Microscope 18, 41-50 (1970).

10. Noritake, C. S., et al., "Polarized light brings out details of fracture zones," Metal Progress 99, 95-8 (1971).

11. McCrone, W. C., "Ultramicrominiaturization of microchemical tests," The Microscope 19, 235 (1971).

12. McGraw, H. R., "A microscopical method for measuring amplitudes and rates of vibration," The Microscope 20, 369 (1972).

13. Richardson, J. H., Optical Microscopy for the Materials Sciences, Marcel Dekker, New York, 1971.

14. Forlini, L. and W. C. McCrone, "Dispersion staining of fibers," The Microscope 19, 243-54 (1971).

15 McLean, J. H., "Interference microscope birefringence of fibers," <u>Textile Res. J.</u> 41, 90 (1971).

16. Tolansky, S., <u>Multiple-Beam Interference Microscopy of Metals</u>, Academic Press, New York, 1970.

17. Dehoff, R. T. and F. N. Rhines, <u>Quantitative Microscopy</u>, McGraw-Hill Book Company, New York, 1968.

18. Ruch, F. and L. Trapp, <u>A Microscope Fluorometer</u>, Zeiss Info. No. 81, 1972/73.

19. Brown, K. M. and W. C. McCrone, "Dispersion staining, Part I," <u>The Microscope and Crystal Front</u> 13, No. 11 (March-April 1963).

20. Brown, K. M., W. C. McCrone, R. Kuhn and L. Forlini, "Dispersion staining, Part II," <u>The Microscope and Crystal Front</u> 14, No. 2 (September-October 1963).
Both of these references provide excellent coverage of the theory and application of dispersion staining, including many illustrations and photomicrographs.

21. Tausch, W., <u>Fundamentals of Fluorescence Measurements</u>, p. 111, Zeiss Info. No. 54, 1964.

22. Gahn, J., <u>Polarized-Light Microscopy by Transmitted Light</u>, p. 89, Zeiss Info. No. 61, 1966.

23. Torge, R., <u>The Interference Microscope</u>, p. 100, Zeiss Info., No. 61, 1966.

24. Dimersoy, S., <u>The Development of the Microscope-Photometer, with Special Reference to Reflectivity</u>, <u>Zeiss Mitteilungen</u> 4, No. 6, 254-279 (1967).

25. Lang, L., <u>Nomarski Differential Interference Contrast Microscopy, Part IV, Applications</u>, p. 22, Zeiss Info. No. 77/78, 1971.

CHAPTER 5

COUNTING AND IMAGE ANALYSIS

A. COUNTING ANALYSIS

1. Introduction: Counting is that process whereby a one-to-one correspondence between a number system and items to be counted is accomplished; in other words, to indicate or name by units or groups to find the total number of units involved. For a very few items, or units, the process of counting is simple and straightforward with a low confusion factor. Two factors that make counting difficult for people are quantity and variety. Quantity involves memory as to how many units were counted before, and variety involves differentiation of units to be counted from other units of a different size, shape or color. The combination of the two factors is common in the counting of microscopical objects.

The process of counting is often purposeful in the enumeration of individual components possessed of some interesting attributes, life or death, infested or free, red or white etc. It is also involved in measuring some continuous variate like yield, diameter, length, assimilation, growth etc.

Mixtures in which an undue proportion of naturally associated impurities may be present such as paper pulp, fabrics, felts, ore concentrates, refractories, rocks, alloys, explosives etc. are very susceptible to being characterized by counting methods. The characterization of the latter type of material would be that of "quality" as compared to allowable standards of percentage impurities.

Since the end result of counting, to find a total number of units, is usually in microscopical work only one step in the determination of other information (such as area, volume, weight and other physical constants or attributes of the material under study, and many of the processes for such determinations rely on large numbers counted, to be totalled and perhaps even treated statistically) it is important that the counting be as accurate as possible, and done rapidly to conserve time.

Much of the following is devoted to the brief treatment of counting processes for characterization of sample materials. Because of the tedium and inherent errors in the manual processes, automatic methods of counting and characterization are used more frequently in all phases of microscopical work. Technology has recently provided the means to greatly increase the volume and accuracy of such work many times over that of manual methods. Although automation in this area is very complex and is becoming a field in itself, it is appropriate to include at least a brief survey of such methods later in this chapter.

2. Sampling: As a rule, a sample is interesting only insofar as it furnishes information about the population from which it came. In counting operations under the microscope we may be concerned in counting quantity only, or attributes only, of a particular population. The larger the sample, the more nearly its attributes, numbers etc., are likely to approximate the population probability.

In particle counting etc. the process of sampling is very important to the accuracy of the results. Generally, the process of sampling is divided into three major operations:

(1) Collection of a gross sample.
(2) Reduction of gross sample for transportation to examining location.
(3) Preparation of sample for analysis.

The methods of sampling vary widely, depending upon the material to be examined, but for any particular type of investigation sampling techniques should be standardized, especially where comparisons are involved. For instance, in collecting a blood sample from a finger, unless an incision is made 3 to 4 mm deep, allowing for the free flow of 3 or 4 drops without pressure, the white cell count may be grossly inaccurate and worthless.

In cases where the sample is collected remote from laboratory facilities its treatment and handling must be performed in a known manner, preferably standardized, to prevent contamination or changes during transportation and/or storage.

In preparation for microscopical examination particular attention must be paid to cleanliness, proper diluents if used and the use of effective dispersants to improve accuracy of counts.

3. Aids to counting

a. Reticles: Many reticle patterns devised both for general and for specific uses are designed to reduce the confusion factors mentioned previously. The crossline, squared grid and other simple geometric patterns are used to divide the field into areas which are limited by the pattern lines, thus reducing the count per area, each area designated acting as a small "memory" to store already counted units. For counting components of greatly varying densities in a specimen, two squares can be used, a large one with a small included one. The frequency of the sparsely distributed particles in the major square, for instance, is related, in the proportion of the ratios of the area of the small square to the larger, to that of the denser population counted only in the smaller square.

Other patterns serve to delimit the field of view to specific areas and thereby the delimiting lines serve as "reminders" that the area outside is not to be counted, reducing confusion, or conversely that the area "included" is only to be counted. Reticles which "mask" areas serve much the same purpose. Specific areas of special shape, either opaque or transparent, are used in reticle patterns as constant "reminders" of comparison to the viewer, and thereby reduce the effort of differentiation. When they can be made of specific size, they also can serve as measuring aids.

 b. Counting chambers: To provide assurance that the same quantity (sample) of material is taken for each count, counting chambers are sometimes used.

Counting chambers have been devised for a number of specialized purposes. Determination of the constituents of blood, such as red and white cells, by percentage volume; numbers of microorganisms by volume in water examinations etc. are two common applications.

All counting chambers have one attribute in common, that of a chamber of known depth, and sometimes of specific volume. With a known depth between the bottom of a reservoir to be filled with the material to be examined and the underside of an applied coverslip, volumes are easily measured. Sometimes the coverslip itself is ruled in specific area intervals for the purpose. In other cases, an ocular reticle with an appropriate pattern is used. Some chambers are filled before the coverslip is applied, and some have special "charging inclines" that allow filling after the coverslip is applied. In blood counting, the latter type of chamber improves the accuracy of the count. The chambers are usually furnished with "overflow moats" that assure complete filling of the void between the base and underside of the coverslip.

To provide practical information relative to dimensions, cavity or chamber sizes, markings and other features of counting chambers, a brief description of some of the more common types follows. It is intended that they convey an order of magnitude and variety of form to the microscopist that will be useful in his selection and ultimate use of some of them.

COUNTING ANALYSIS

(1) <u>Counting chamber, bacteria (Petroff-Hauser)</u>: Open parallel cavity cell, Neubauer ruling. Cell depth 0.02 mm. Glass slide is 43 x 42 mm. Mounted in a bakelite holder 75 x 45 mm so arranged that the underside of the slide lies within 0.1 mm of the plane of the microscope stage.

(2) <u>Counting chamber, bacteria (Helber)</u>: Same chamber dimensions as Petroff-Hauser. Ruling is 1/400 square mm., improved Neubauer pattern. Slide dimension is 75 x 35 mm, open-moat, one-piece construction, 5 mm thick. The latter limits this slide to stained or more easily recognized specimen material. Critical focusing might not be possible.

(3) <u>Counting cell, bacteria (Dunn)</u>: This accessory consists of two pieces, all glass. The bottom plate or slide is 37 mm x 60 mm x 2 mm thick and has polished surfaces. A top plate with a fine ground surface is furnished that has dimensions of 38 x 38 x 1 mm thick and a 20 mm diameter center opening. In use, the top plate is placed on the polished bottom plate and a tight seal is obtained between the two parts by wetting their mating surfaces. An ordinary microscope coverslip may be used as a cell cover. As the glass is of a heat resistant type, this cell will withstand repeated sterilization and the use of acid cleaning solutions.

(4) <u>Counting chamber (Howard)</u>: For counting procedures in suspension emulsions and powders which are brought into suspension. The chamber is of one-piece construction. In the center of the plate is a polished platform counting area of 15 mm x 20 mm, surrounded by a moat and flanked by shelves on both sides to support the coverslip which has been polished optically smooth. There is an accessory micrometer reticle ruled into squares useful for mold counting. It has a large square 9.96 mm on a side ruled into 36 squares, each with an area equal to one-sixth the diameter of the opening in a 10X Huygens ocular diaphram, with the microscope properly calibrated for the Howard method. A coverslip 33 mm x 1.0 mm thick is furnished with the chamber

There are a number of variations of the Howard counting cell. One is a special slide with a circular well. In the center of the well there is a stage or platform which is 0.100 mm lower than the surface of the slide. When any liquid is placed on the surface of this central platform and covered with a plane coverslip, the depth of the liquid is exactly 0.100 mm. Any excess liquid will overflow into the moat which surrounds the stage. The coverslip used is rather thick and polished.

In another version of this very popular counting cell, a glass slide 76 mm x 35 mm with a central circular well 0.1 mm deep is used. A special coverslip which defines 25 calibrated fields, each 1.382 mm in diameter, is available. This obviates the necessity for adjustment of microscope and ocular reticle as required otherwise.

(5) <u>Counting chamber, blood</u>: There are so many variations of this most important counting chamber that only a few of the pertinent and/or interesting details will be included here. Some features that have been devised to make blood counting more accurate and rapid are:

(a) "Bright-line" rulings for better visibility.

(b) Double-chambers to permit counting of red and white blood corpuscles at the same time without the necessity of cleaning or refilling.

(c) "Chamber-charging incline" which eliminates overloading and acts as a reservoir. The liquid cannot commence its flow until a sufficient quantity to fill the chamber is applied. Then it enters with one swift influx, minimizing the danger of uneven distribution and bubbles caused by a slow halting influx due to an insufficient application.

(d) Quadruple chambers to permit making duplicates of both red and white cell counts, for checking purposes, without the necessity of cleaning and refilling the chamber or refocusing the microscope.

Most blood counting cells are 0.1 mm in depth, except some that are designed for counting cells in cerebro-spinal fluid or for leucocyte counting, wherein they are obtainable with chamber depths of 0.2 mm.

(6) <u>Counting chamber, water microscopy (Sedgwick-Rafter)</u>: This chamber is designed primarily for the microscopical examination of water, but also may be used for dust examination as it can be hermetically sealed. It consists essentially of a glass slide 33 mm x 70 mm with a 20 mm x 50 mm chamber 1.0 mm deep. All glass construction includes cover supports inlaid in the base. A coverslip 0.5 mm thick, optically plane on both sides is ordinarily used.

A Whipple ocular micrometer disc, consisting of a network pattern of a large square 7.0 mm on a side subdivided into four smaller ones, each of which is subdivided further into 25 small squares, is quite often used with this counting chamber (Figure 88).

COUNTING ANALYSIS

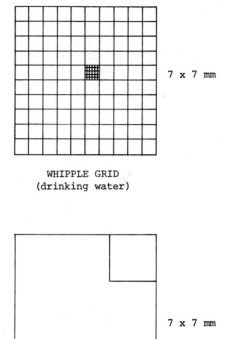

WHIPPLE GRID
(drinking water)

MILLER SQUARES

Figure 88. Counting reticles. (adapted from illustrations furnished by Graticles Ltd.)

c. <u>Dispersants</u>: In counting various components of a suspension, it is important that the components do not clump or aggregate. For the sample under examination to best represent the whole and for ease of counting, the units to be counted should be randomly dispersed and not so numerous as to make counting difficult.

Known dilutions of the sample to be counted make it simple to express the count in standard units of one kind or another, cubic centimeters, liters etc. The diluting fluid may be a good dispersing medium or it may not. Dependent upon the nature of the particles of material to be counted, various fluids may cause them to clump or form specific formations that are detrimental to the counting process, and ultimately perhaps to a statistical analysis. In some cases it may be necessary to experiment with various dispersing agents to attain good results. Some of the fluids for the purpose are:

Water	Acetone
Water and glycerine (1:1)	Cyclohexanol
Water and ammonia	Butanol
Water and potassium citrate (2%)	Dammar
Water and Calgon (0.1%)	Turpentine
Glycol	Duco Cement
Glycerine	Glucose syrup
Ethanol	Xylene
Polymethylmethacrylate	Polystyrene
Solutions of surface active agents	

In blood counting, special standard diluents and dispersants are used.

d. Fractionating specimen material

(1) Sieves: Before the particles of material to be counted are conditioned for microscopical examination, it may be desirable to isolate sizes or divide the constituents into known fractions according to size. With granular and powdered material this is normally accomplished with sieves. Both dry and wet screening are used. For gross and macroscopic samples, standard sieves are available and the finer ones are quite useful in microscopical separations. The sieve openings range from 4760 μm down to 44 μm. This range is covered by U.S. Standard series sieve numbers 4 through 325, comprising 28 different size openings. The tolerance in the size of the openings (ranging from $\pm 3\%$ to $\pm 8\%$), and the fact that particles are three dimensional and the openings of sieves are areas having two dimensions only makes sieving methods of separation at best an approximation. However, this type of preparation prior to counting particles can reduce confusion and time for counting, especially where size ranges of desired particles for counting are known beforehand.

In the lower micrometer range, below 44, the standard construction sieves (either woven wire or perforated plates) are subject to severe accuracy limitations. Buckbee-Mears Company furnishes a complete line of sieves of electroformed construction. The sieve material is a planar nickel electroformed surface with a hole size range certified to a tolerance of $\pm 2\ \mu m$. Hole sizes are available from 90 μm downward and including 5 μm. The electroformed construction, in addition to improving accuracy, is easily cleaned and free of particle entrapment that occurs in woven meshes.

(2) <u>Heavy liquids</u>: Heavy liquids are commonly used by the mineralogist to assist in identifying minerals by specific gravity. Separation of particles by the same means is a useful way to reduce the effort in the counting process. If it is known that the particle matter to be counted is of a certain specific gravity, it may be isolated on that basis. Some of the so-called heavy liquids for specific gravity determination or for separation purposes include carbon tetrachloride, solutions of mercuric iodide in potassium iodide (Sonstadt solution), borotungstate of cadmium (Klein solution) etc. Many other solutions can be prepared for this purpose. Commercially available solutions are prepared in a series of specific gravity steps that are very useful. Briefly, the material is introduced into a solution of high specific gravity and the solution diluted until one constituent after another sinks and is removed.

(3) <u>Filters</u>: Material to be counted can be separated by filtering, the material desired either passing through or remaining on the filter, as the case may be. Very accurate paper and/or plastic filters are available commercially for the purpose.

(4) <u>Ultrasonic devices</u>: In recent years a number of devices utilizing ultrahigh frequency vibrations have been developed for cleaning, mixing and separation of fine particles and suspensions.

For example, an "ultrasonic sifter" is available to perform dry sieving operations on particles smaller than about 37 μm to as small as 5 μm. The range of material particles that the device can separate is from 850 μm to the 5 μm size mentioned above.

It operates by a vertical oscillating column of air plus a repetitive mechanical pulse. The column of air is maintained in an oscillating state by a flexible diaphram operating to move the air column in the sieve stack. A mechanical pulse is applied to the sieve stack to reduce sieve binding and to continuously rearrange any clustered particles formed by agglomeration. The finer sieves used in such devices, instead of being woven wire, are usually electroformed meshes (Figure 89).

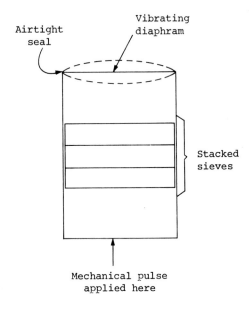

Figure 89. Ultrasonic sifter. (adapted from illustrations courtesy ATM Corporation, Sonic Sifter Division.)

Dispersal of particles is also achievable by supersonic devices principally designed for cleaning and/or mixing. They will provide excellent small particle suspensions of clays, carbon blacks and soil colloids. Intelligent use of such devices can improve accuracy of counting analysis with the microscope, and for many routine or large scale studies is a valuable time saver.

(5) <u>Micropipettes</u>: For precise sample intake and dispensing, with high reproducibility, commercially available pipettes can be of great assistance in certain counting analysis work. Fixed volume pipettes are available from 2 through about 50 μl. Variable units also can be had for volumes from 2-10 μl. Volumes are reproducible to within ±3%. Such pipettes are usually very well suited to blood cell counting techniques.

COUNTING ANALYSIS

4. <u>Counting procedures</u>: Once the sample has been selected, dispersed and/or otherwise prepared and ready for actual counting under the microscope, certain procedures are necessary for best accuracy and conserving of time.

    a. <u>Form of tabulation</u>: The format of the record of counted units will vary, of course, depending upon the use of the count. Preformed tables already ruled and divided according to results desired should be prepared beforehand. Adaption of business and accounting forms for this purpose is excellent practice, as numerous configurations of columns, sizes and widths are in plentiful supply. Also, the paper is sturdy and will take many erasures without its surface being damaged. Pencil entries are advisable.

As an example of a planned format and layout for recording, consider Figure 90. Note that the number of fields to be counted is ten and the number of elements in each field counted is ten.

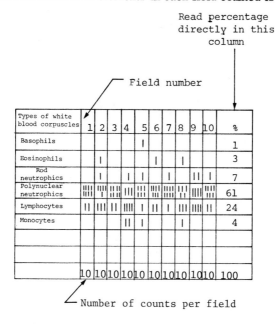

Figure 90. Counting chart for differentiating white blood corpuscles.

With this arrangement the number of items counted then is 100 and if the total of each type is recorded at the right in a separate column, the percentage of each element of the count is read directly, no computations being necessary. Considerations of the number of fields to be counted, the number of counts per field to be made, and the ultimate use of the count can often be used to devise a tabulation system that records the count, and at the same time provides time-saving assistance for obtaining the results desired.

If statistical studies are to be the ultimate use of the count, then appropriate forms for counting for that purpose should be used. Elementary texts on statistical methods indicate formats suitable for frequency distributions and other systematic arrangements of data collection.

The emphasis on tabulation is made here because in the usual course of performing manual counts for analysis purposes, the number of counts to be made are in the hundreds or even thousands. Any way at all of conserving time or exploiting the count for maximum information is very worthwhile.

b. Counting fields: Counting microscopical subjects is performed by "fields", usually coincident with the field of view at the ocular. The ocular field of view is circular and quite often for counting purposes is divided into square or rectangular areas by special coverslip patterns, as with blood or other specialized counting chambers, or by reticles in the ocular itself.

Special counting oculars are also available. Zeiss provides one that contains two diaphram blades with rectangular cutouts which can be shifted symmetrically in relation to the vertical center line, thus making it possible to select square sections from the field of view. When using an object chamber of known depth, volumes can thus be easily limited for counting purposes. They also make available a special ocular with adjustable counting lines. Primarily developed for plankton counting, it is also useful for other purposes. There is a fixed horizontal line and two vertical lines, each of which can be shifted symmetrically in relation to the center of the field by means of a knurled ring. The two adjustable lines can then be used to delimit counting areas of various sizes.

c. Magnitude of the count: Just how many items should be counted is very dependent upon the purpose of the counting operation, the nature of the counted elements and how they were collected and prepared. If the count is to be compared with

COUNTING ANALYSIS

standards already established (per unit area, volume etc.), 10 or 20 fields (ocular fields) are usually sufficient, as in the case of blood counts, and the total element count may be in the vicinity of 100 or so.

If, on the other hand, and this is the rule rather than the exception, the count is to be used in a statistical way, the number of items counted is much higher. At least two or three times that for the former condition. Counts of 200 to 300, or a thousand or more, are quite common per "population" in work relative to frequency distributions etc.

The ultimate determination of whether the magnitude of the count is sufficient will necessarily rest with the investigator and his knowledge of the subject. In comparison studies this is easily decided. In statistical studies there are special means available to test whether the "population" has been sampled properly. Some knowledge of statistical methods is necessary for those determinations.

If the items to be counted are regularly shaped or easily recognized forms, a count of 300-400 is usually sufficient to be representative of the population. With this count, accuracies of about ±5% will be experienced. Counts of 400-800 will improve accuracy to about 1 to 2%.

When counting irregularly shape particles, one of the major difficulties is associated with the subjectiveness of sizing them. As standard methods for assessing particle size show, errors of evaluation of the population are associated with the size of the sample examined (the number counted). This error is inversely proportional to the total count. Thus, for an analysis to be representative it is necessary to count upwards of 1000 for an average sample, and in cases where there is a wide range of sizes it may be necessary that the count be considerably higher.

Assumption that at least 300 counts will be necessary, for instance, may serve as a guide to the size of the "population" to be included in the count. For very small particles of high concentration, a portion of the material included under a single coverslip may be sufficient. If the items are comparatively large and/or are widely dispersed among other elements of the "population", then the area included by an entire coverslip of even large size may not be sufficient. In any event, some knowledge of the material either by past experience or by a preliminary examination will provide the investigator with information such that he can decide on the size of coverslip to use and/or how many microslides might be necessary per count.

d. Search patterns: Counting items within a specific area depends to some extent upon the shape of the area to be covered and the availability of the countable elements. The area can be covered in a systematic way, assuming that the items to be counted are dispersed in a random fashion. Systematic search of a circular, square or rectangular area may be performed in any one of a number of configurations. If the available countable elements are very plentiful (many times the required count) the search need not be over the complete area, but need be only composed of a sufficient number of fields as to accommodate the maximum count required.

If this latter condition holds, advantages are twofold. One, the danger of double-counting between successive ocular fields of view can be eliminated by spacing the fields sufficiently and, two, the fields can be located away from any possible localized clumping or nonrandom distribution particle dispersal points. In spite of all precautions that can be taken in preparing such slides for counting purposes, the actual mechanics of doing so quite often result in localized nonrandom and aggregated portions of the "population" in certain areas under the coverslip. For circular coverslips these conditions often exist at the edges. In squares and rectangles, the corners and edges are points at which the conditions of clumping, aggregation etc. occur. These areas should be avoided in counting, to assure a more representative count within the population.

For circular coverslips the center can be marked temporarily with a dot of Indian ink. This is located under low power and the stage coordinates noted. Draw a circle denoting the coverslip on a piece of paper and by knowledge of its diameter and the located center in terms of stage coordinates, lay out a scheme of counting fields. Figure 91 illustrates the method.

Square or rectangular areas can be similarly laid out to assure a representative sampling during the counting process. As long as the total elements to be counted greatly exceed the number of counts to be made, sufficient spacing of the counting fields to preclude double-counting is simple and they can be widely separated over the area for that reason.

When the density of the elements to be counted is such that their number may be only somewhat greater than the total count desired, a nearly complete coverage search will be necessary. It is under this circumstance that care should be taken in not allowing ocular fields of view to overlap in either the x or y di-

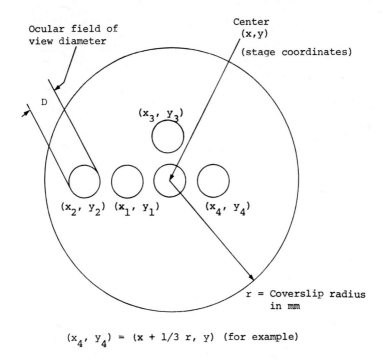

Figure 91. Counting-field plan for circular cover.

rections. Figure 92 illustrates a nearly complete search scheme with the precaution accounted for. The minimum spacing of the ocular fields of view along the search track will be one field diameter. The spacing between tracks is the same. When there is sufficient countable material, the spacings may be increased appropriately.

Knowledge of the diameter of the field of view in millimeters provides information for setting up the spacings. For instance, the field of view of a typical 12.5X Huygenian ocular used with a 40X objective is 0.31 mm. This dimension is of sufficient magnitude as to be easily set up on a graduated mechanical stage as, almost without exception, such scales can be read to the nearest 0.1 mm. The field of view diameter is given by manufacturers

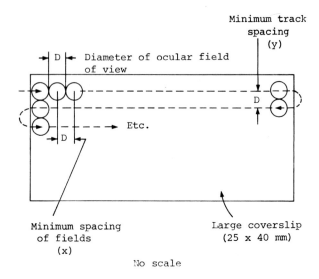

Figure 92. Counting field plan
Rectangular cover.

for various combinations of oculars and objectives, or they supply a "field of view number" which is used to compute the field of view diameter with a given objective magnification. This number or index is a measure, for a given ocular, that is indicative of how much of the intermediate image, which is limited by the rim of the ocular tube, can in fact be observed. The effective diameter in mm of the intermediate image is expressed by this number or index. With binocular observation the interpupillary distance of the observer's eyes limits the diameter to about 30 mm. The field of view diameter can also be determined by viewing a stage micrometer with the optics that will be used in the counting process.

Referring to Figure 92, the count is started in one corner of the "population field" or coverslip in this case, and follows the search scheme indicated by the arrows. Each time a track is completed, the mechanical stage is set for one field of view diameter in the Y direction and continued in a horizontal track search to the opposite vertical edge etc. If the work is interrupted, the notation of the mechanical stage coordinates and the direction of the track search at that point should be noted to pre-

COUNTING ANALYSIS

vent any indecision and possible double-counting when the work is again taken up. Each track can be set up beforehand on the recording form with an "edge index" comprised of the Y setting of the first track plus the field of view diameter successively for each track thereafter. This nearly-complete search means just that. It will be noted that not all countable elements present can be accounted for in this scheme, as the circular fields of view do not cover the interstitial spaces. If it is desired to count every countable element, then square or rectangular areas of view must be sequentially examined. This can be done with the aid of net patterns in the ocular.

(1) <u>Double counting</u>: Counting within the ocular field of view can be done with or without reticles. If the items to be counted are few in number, no reticle needs to be used. If there are more than a few, a reticle which divides the field of view into four or more parts is very helpful in reducing confusion and the danger of double counting.

A consistent approach as to what is to be counted relative to the vertical and horizontal lines of the reticle, or to the circular edge of the field of view, is necessary for greatest accuracy and avoidance of double counts. Reference is made to Figure 93. The conventions adopted might include counting only items which are wholly within the field, and none which touch or are obscured by the field edge. If the items are in a recognizable "whole" form, then one should include the precaution to count only "whole" items and those that are at least half-size or larger, disregarding smaller fragments. This is often the case, for instance, in counting microfossils and entails recognition of what is "whole" and what is not. If this precaution is not taken, errors will be introduced by including in the count two halves, or many particles, of the same item.

When the reticle is in place, in order to protect against double counting from one subdivision of it to another, a convention is necessary. For instance, count only items which are within the subdivisions and which touch the left-side vertical and lower horizontal. Items touching the right-side vertical and upper horizontal are not counted.

5. <u>Slide preparation</u>: When specific purpose counting chamber slides are used, as in blood counts, water analysis etc., the preparation of the material is quite standardized and the material is in such a form as to be used once in the counting process and then discarded. The chambers are, of course, used over and

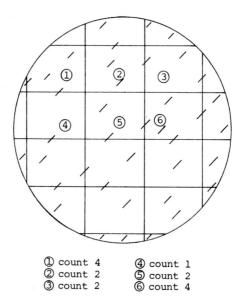

Figure 93. Example of counting by convention within a field.

over again.

For the multitude of other types of counting analysis work to be performed, the slide preparation may be either a temporary or permanent mount. Many of the same procedures and techniques briefly covered in a previous chapter are employed in their preparation. However, for the counting process, there are a few matters, if taken into account, that will assist in differentiation, ease and accuracy of counting.

Previous mention has already been made of the use of proper

COUNTING ANALYSIS 249

diluents and dispersing agents. That is important in getting the countable material into a suspension that is as representative, frequency-distributionwise, of the sample population as possible. Getting the material so dispersed into a position under the microscope is quite another matter.

(1) Mounting technique: Temporary slide preparations may be made using counting chambers. Even though they are designed for specific uses, most of them can be used for many other counting purposes as well, especially where volumes are a factor in the analysis. It is just as appropriate to determine the percentage by volume of many substances by using a blood-counting chamber as it is to determine a blood constituent percentage by volume. The same attributes of the chamber, known depth and ruled coverslip areas, are just as useful for one purpose as the other. Of course, the material particles must be of the same order of size magnitude and susceptible to being counted in a liquid suspension.

Temporary mounts in fluids or resins are made, of course, using standard size microslides and coverslips. Fluid mounts are made by applying the material, in liquid suspension, to a microslide and covering. Application of the liquid suspension to the slide surface by dropping it from a pipette held a few inches above the microslide will assist in adequate random dispersal of the particles over the area to be covered. Dispersal may also be aided by a circular motion of the coverslip with the eraser end of a pencil. If the fluid is of such a specific gravity as to allow the particles to settle onto the slide surface, they will do so in a well distributed random manner. If the suspension fluid has a relatively high specific gravity, the particles may remain partly in suspension and partly on the surface of the slide. If the liquid layer is not maintained thin enough, counting difficulties will be encountered, as particles will perhaps be in and out of focus in different levels of the suspension. Also, if the suspension fluid has not been selected properly, this tendency to "float" particles can result in serious counting errors.

The best method for making slides for counting purposes is with resinous media. The material suspended in a fluid is dropped by pipette not on the microslide as before but on the coverslip. The fluid is driven off by gentle heating so as not to clump the drying particles. A suitable quantity of resinous media is then applied to the microslide which is used to "pick up" the coverslip. The assembly is then inverted and gentle heat (if necessary) is applied to drive off the solvent. Carefully done, this procedure will result in a sample of the material being distributed in a single layer on the undersurface of the coverslip, an ideal situation for statistical counting especially.

To assure the adhesion of the particles to the coverslip surface during "inverting" and further heating to drive off volatiles, it is sometimes an advantage to include a drop or two of gum tragacanth (suspended in water or alcohol) in the fluid suspension, before dropping it on the coverslip.

When recognition of the particles to be counted may be difficult, it sometimes can be assisted by proper selection of the mounting medium. A particular refractive index medium, for instance, will render certain particles not to be counted faint and transparent, thus facilitating differentiation. The converse is true, of course. The choice is dependent on the nature of the suspension and the qualities of the countable particles.

Chemical differentiation of countable items is also possible. The application of certain stains will provide color differences. Other chemicals which act in a known manner can differentiate by changes in shape (crystallization), color and/or precipitate out of solution undersired factors in the counting process. The use of a hot stage to create certain observable characteristics typical of countable particles or the use of polarized light or fluorescence techniques can be very useful in differentiation also.

6. <u>Compositional and quality counts</u>: Comparisons of component particle counts against the properties of materials of known or standard composition are often used as a measure of quality.

The count of impurities such as seed hairs or bran particles in flour, hulls in rice bran, organisms in water, molds in tomato products etc. can result in quality-level determinations. By comparing with similar counts on mixtures having known impurity content, a comparative quality figure is obtained. Certain standards of quality will sometimes have been established that permit the same type of comparison.

In water quality analysis, counts of various organisms, both plant and animal, are made in sufficient quantity to determine the units of each in a cubic centimeter. The tabulated results are compared with standards previously established.

A ratio method of determination of composition is to use a reference substance such as lycopodium powder. It is uniform in size, averaging 94,000 spores per milligram, and its appearance is unaffected by triturating in a mortar or boiling in dilute acids or alkalies. A definite measured amount is added to an unknown and an equal amount to a sample having a known composition. Without any precautions as regards the measurement of the preparation, counts of particles in question and of the reference substance are made. The ratio of the counts is dependent upon the composition of the sample being analyzed.

A standard set of suspensions of 10, 20, 30% etc., may be used to establish points on a curve for which the count of the number of particles in each percentage case is a point. Then unknown suspensions can be counted under the same conditions and the percentage curve used to determine the percentage count of the unknowns.

These methods, if performed with care using suspensions with well distributed particles, will yield results with an accuracy of 1 or 2%.

Counting techniques based on comparisons against standard charts have been used as a quality control tool for a considerable number of years. Typical of these are methods for determining the quality of iron and steel. The images in the field of the microscope are compared against a standard chart or figure and direct comparisons of certain types of inclusions and/or grain sizes are made on a countable basis. This method, as others of the type, are subject to errors due to a variation in level of assessment that observers establish between the standard chart and the microscopical field. Also, any variation in techniques used when selecting and classifying individual fields during a count will contribute to an overall error. Mostly for convenience, and perhaps for more accurate determinations, these types of "charts" are often incorporated as special reticles in the ocular of the microscope. Some are arranged such that the field of view at the ocular includes the material under study at the center, surrounded by sectors of various standard charts (Figure 94). More elaborate oculars which have a micrometer-disc turret into which various discs are accommodated and which can be rotated into the field of view for comparison with the specimen are also available.

7. <u>Areal and linear analysis</u>: If a microscopic section is taken through an aggregate of heterogenous material, the intercepts of the various ingredients in the plane of the section or in a line across it correspond to their respective percentages by volumes.

To simplify such methods, the specimen should be in compact

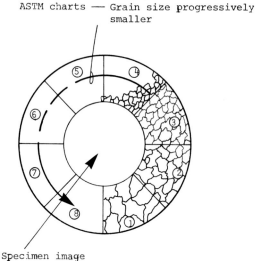

Figure 94. Grain size ocular reticle.

form so that it can be sectioned by cutting or grinding. Alloys, rocks, ceramics etc. are suited to such treatment. Powdered material can be embedded in plastics or other cementing materials, and then surfaced by grinding for measurement.

Camera lucida drawings on coordinate paper or photomicrogaphs of a number of fields can be used in the determination, or percentage areas can be measured directly or indirectly by weighing paper cutouts of the constituents. The percentages then are compared with standards.

With linear analysis, the use of drawings or photomicrographs as above is also used to measure the length of the ingredients. The lengths are put on separate strips of paper for each of the constituents and the totals measured and compared. The total distance measured should be at least 100 to 200 times the diameter (or length) of the average size grains.

The volume percentage furnished by this means of areal or linear analysis will represent percent composition by weight only in case the specific gravities of the ingredients are practically identical. Otherwise, the volume percentage of each component must be multiplied by its specific gravity and the resulting ratios converted to a weight percentage.

It is possible, for instance, using this method to ascertain the percent of a given element (in an alloy, for instance) such as oxygen or carbon, which is present in combination as an oxide, carbide or a eutectic of definite composition. In such a case, the volume percent of the compound is multiplied by the percent of the required element contained in it.

8. Point counting: When the quantitative relationship of certain particles or components to the total volume of a material is of interest, the point counting method is useful and much more rapid than the older preceding methods. In microscopic specimen materials individual components are usually seen as irregular surface elements distributed over the field of view in random fashion.

For the volumetric share of various components it is sufficient to find the ratio of their surface area to the total area. This is especially true when a layer of material is thin and a constant thickness.

The point counting method is basically a simple one:

(1) A point grid is placed over the preparation.
(2) A small surface element belongs to each grid point.
(3) It is only necessary to determine the number of grid points which coincide (hit) with the desired elements to be counted to obtain an approximate value of the sum of the areas of the desired elements. All that need be done to obtain an approximate value of the ratio of areas, then, is to divide the number of hits by the total number of points. The accuracy of the approximation is increased the finer the mesh (greater number of points) used.

If a grid of points is rotated or displaced, the hit number becomes slightly different and, therefore, the approximate value of the ratios between the areas changes slightly. Instead of a disadvantage, as might be assumed at first, the individual results with different positions of the grid can be averaged and the mean-value accuracy increased thereby.

Because repeating the measurement at different positions of the grid actually changes the grid pattern (relative to the particle distribution), it is unnecessary to use a regular grid pattern. All that is actually required is that the preparation be covered with points in such a manner that no area has a concentration of points anywhere. The average is calculated solely from the total number of hits and the total number of points. For the end result it makes no difference whatsoever how many hits are obtained at any particular grid position.

It can be shown that the ratio of the areas lying within the visual field of the ocular should be, as nearly as possible, the same as that in the whole preparation. Experience has shown that this is the case when the average diameter of the particle to be measured is about equal to the distance between two grid points. With particles magnified to such a size, the hits can be generally established with certainty. This means that there is a *best* magnification to use with a given grid interval, and if for some reason the magnification cannot be changed then the grid interval should be, to fulfill the condition stated previously for best accuracy (Figure 95).

Because of the relative ease of point counting and its importance to counting analysis a brief treatment, from counting through error estimation, will be provided herein.

First, a reticle of points is installed in the ocular of the microscope. For the example here we will use the one illustrated in Figure 95 of 88 points. A series of *settings* is now made and the number of *hits* recorded for each setting. The first setting may be such as in Figure 95 in which 17 hits are recorded. The next setting may either be made by rotating the ocular points to a new relationship with the same particles, or by moving the specimen with the mechanical stage to a new group of particles. It makes no difference which is done. Likely, a combination of rotating the grid points and moving the specimen material throughout the analysis will be a common method. In any event, at each *setting* the hits are counted and tabulated. The total number of hits N_p is obtained by adding all hits (by whatever method they are obtained). Multiplying the number of *settings* by the number of grid points per field (in this case 88) furnishes the total number of points N. The ratio of N_p/N, then, is a point-hit-count ratio that represents the ratio of the actual areas of the counted particles $(F_1+F_2+F_3+\ldots\ldots)$ to the total area covered (F) or F_p/F.

This point-count ratio is an approximation, then, of the ratio of areas to be determined. The greater the number N, the closer the approximation and the greater the accuracy of the count. However, the microscopist will want to know how close the approximation is or how much the error might amount to. Referring to Figure 96, the idealized bell-shaped curve repre-

COUNTING ANALYSIS

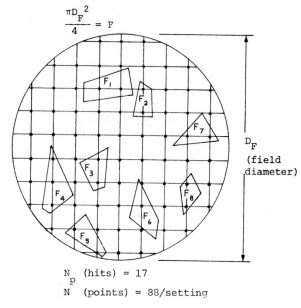

Figure 95. Point counting
Point spacing vs. particle size per field.

sents graphically the probability distribution of how closely the approximation of the point-hit-count ratio to the actual area ratio can be determined. An equation has been developed which relates the error limits on a probability basis. The equation is

$$\delta = 0.674 \sqrt{\frac{p \cdot (1-p)}{N}}$$

where $p = \dfrac{F_p}{F}$.

δ is called the <u>confidence limit</u>. It denotes the upper and lower limits within which the ratio N_p/N will approach F_p/F with a probability of 50%. At double the value of δ the probability is 82% and at triple the value of δ the probability is 96% (considered as good as a certainty). The percentage of probability is termed

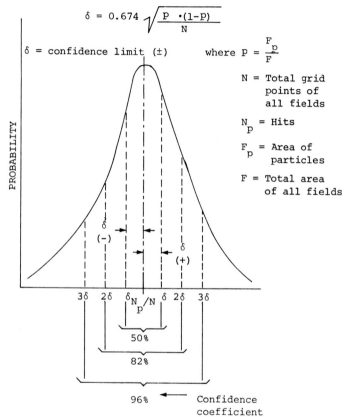

Figure 96. Point counting. Estimation of error.

termed the <u>confidence coefficient.</u>

Assuming that the total number of hits N_p is 1700 and the total number of points N is 8800, N_p/N is 1700 divided by 8800 or 0.193. The confidence limit is

$$0.674\sqrt{\frac{0.193\,(1-0.193)}{8800}}$$

or approximately ±0.0028. This means, then, that the ratio N_p/N of 0.193 is correctly representative of the actual area ratio F_p/F within the limits of 0.193 ±0.0028 or between 0.1902 and 0.1958 on a probability of 50% and for a probability of 82% (2 δ) is correct within the limits (0.193-0.0056) or 0.1874 and (0.193 +

COUNTING ANALYSIS   257

0.0056) or 0.1986. For a practical certainty (96%) (3 δ) the value is correct within the limits 0.1846 and 0.2014.

Now, if after these determinations, the limits for a practical certainty (96% probability), for instance, do not fall within the requirements for accuracy for other calculations based on these figures, the error limits can be reduced for any probability by increasing the total number of points N. The equation, however, does indicate that the confidence limits vary inversely as the square root of the number N.

As an example of what this means, if it were desired to retain the same limits of error indicated for a 50% probability and increase the confidence coefficient to 96% (or a practical certainty) for those limits, the total count would have to be 9 times 8800 or 79,200! Thus, it can be seen that the setting of realistic limits is very important from the standpoint of time spent in counting and in even, perhaps, determining the feasibility of the method for the results desired.

Deviations are found to be rather smaller than expected from purely theoretical considerations. The reason for this is probably the fact that when a grid point falls on the edge of a particle, the microscopist will rather count it as a hit than not count it. In doing work of this type, it is important that some "ground rules" be set up by the microscopist regarding what he will, and will not, count as hits prior to beginning.

Practical point-counting reticles may be of various configurations and numbers of points. Zeiss markets an Integrating Micrometer Disk Turret which provides for points numbering 25, 100, 400 and 900. The revolving disc has seven openings. For optimum geometric adaption to the object, integrating discs are mounted in four of these openings. The discs have the same base length. In addition, the large clusters of points are centrally subdivided into 25, 100 and 400 points so that small object fields can be uniformly counted with these partial nets.

Point counting is basically an area ratio measuring technique, and not suited to particle size determinations which require other methods. Nonetheless, within its limitations, it is a powerful and useful tool for the microscopist.

B. <u>PARTICLE GEOMETRY</u>: Particles of all kinds are included in counting analysis techniques in light microscopy. Usually simple visual recognition and tabulation of them is accomplished as with other items of specimen material.

However, there has grown a very large technology that is

associated with the actual character of particles. Some, in specific areas such as the steel industry, relate the type, size and nature of inclusions in iron and steel. In the fields of industrial hygiene, manufacturing, food processing and environmental pollution control, particle characterization and analysis are becoming increasingly important.

The characteristics necessary of determination range from the nature of the particle substance and its possible toxicity to individual particle size and shape relative to their control and/or removal as undesirable elements. Techniques common in chemical microscopy, petrography and spectrographic analysis are all brought to bear on problems of particle identification with the light microscope.

The characteristics of particles that are important in quantification, control and/or removal quite often involve their geometry; that is to say, their dimensional properties including shape. These attributes are measurable and countable and, if obtained in sufficient numbers, can be treated statistically. Results of such studies can serve as reinforcing information to other characterization techniques, or as ends in themselves for quality assessment and mechanical control or removal.

1. Particle size and shape: Because of the generally irregular and random shape and size of particles, it has become necessary to establish artificial standards of geometry in this regard.

Particle sizes are often expressed in diameters which are variously defined. Commonly used diameters are:

(1) Globe and circle: Defined as the diameter of a circle whose area is equal to the projected area of the particle.

(2) Martin's diameter: The distance between opposite sides of the particle, measured crossways of the particle and on a line bisecting the projected area.

(3) Short dimension: The shorter of two dimensions exhibited.

(4) Average dimension: This can be either the average of two or three dimensions exhibited by particle.

(5) Two tangent distance: The distance between two tangents to the particle, measured crosswise of the microscopical field and perpendicular to the tangents.

PARTICLE GEOMETRY

All of the above are statistical diameters in that they are defined for a large number of particles, 200, 300 or more (Figure 97).

Diameters for particles have also been established on their physical (as opposed to geometrical) properties. Sedimentation rates in liquids of specific viscosity are determining factors, as an example.

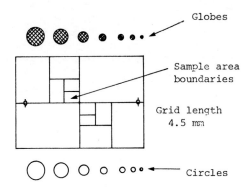

GLOBE AND CIRCLE
British Standard No. 3625

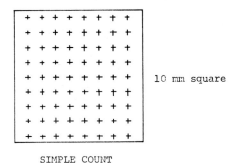

SIMPLE COUNT

Figure 97. Particle size analysis reticles. (adapted from illustration furnished by Graticles Ltd.)

The diameters measured are, in their aggregated totals, converted to statistical mean-diameters. Different ways of computing mean-diameters are available to the investigator, the choice largely dependent upon the purpose to which the particle measurements are directed. The different means include both geometric and physical (weight) inferences.

As aids to particle size analysis, a great number of ocular graticles (reticles) have been devised and are used in conjunction with standards established by scientific and industrial organizations or by governmental bodies in the United States, Great Britain, Germany etc. These graticles include different schemes for globe and circle measurements, point-counting arrays, and length and diameter measurements.

Individual elements of the patterns such as grid squares, dots and circles are calibrated by usual methods employing a stage micrometer. Analysis is achieved by comparing solid particles with globes (opaque circular patterns) and transparent particles with circles (transparent circular patterns). The sample area is defined by rectangles or horizontal bands through which the specimen is racked by the mechanical stage. Some patterns provide accurate and random methods of measuring split-up or irregular areas, and of counting objects of constant size in a multicomponent system of particles.

Assessment of particles as to their dimensions and shape on a manual basis has been and continues to be performed by the light-microscopist investigator, using little more than the techniques and accessories described previously, and much time to accumulate sufficient useful data. The limitations of manual methods have, in almost all cases, been in the lack of rapidity with which such work is accomplished.

C. AUTOMATIC ACCESSORIES: For many physical and mental actions required in measuring, counting and characterization of specimen material with the microscope, there are electrical, mechanical or electronic means to perform or assist in the performance of these actions. In recent years, apparatus ranging from simple mechanical devices to very elaborate electro-mechanical and electronic systems have appeared to assist the microscopist in doing his work more rapidly and/or more accurately. Most of the systems are suitable for attachment or use with conventional light microscopes.

It might be well at this point to mention that the goal of automation is not always to accomplish work faster or more accu-

AUTOMATIC ACCESSORIES

rately, or to reduce the routine and repetitive work of the microscopist. There is an important factor in the utilization of automatic or automated equipment that is at times either an advantage or a disadvantage. That is the removal or reduction of the "human factor". We are all well acquainted with the disadvantages — lack of judgment etc. However, in many types of work with the microscope involving counting and measuring, we are concerned in a statistical way with the data and the end result is required to be a statistical one. Automated recording of sizes, diameters, lengths and widths etc. minimizes the subjective judgments of the human operator and, therefore, provides a better statistical answer. Of course, the parameters are initially set by a human, but during the actual operation the automatic equipment is not influenced by ongoing environmental or thought processes as the human operator might be.

The foregoing is a notable factor in the use of automated means to characterize specimen material, but by far the most important is the ability to accomplish large numbers of single characterizations at rates far beyond the ability of a human observer using manual methods. Statistical approaches to many characterizations rely on large numbers of individual observations of number, size and shape of microscopic particles and/or specimen elements. The greater number of observations and large body of data in this type of work are accomplished from the observational to the computing stage with greater ease when automated systems are used.

Two examples of such systems are covered only briefly below. The field of automation in microscopy is becoming so complex and far reaching this book will merely indicate some typical systems. There are other volumes of this series devoted to the subject in a more detailed manner.

1. <u>Automatic point counter</u>: As an example of a rather simple aid in the accessory category, James Swift & Son Ltd. of England marketed a counting attachment for the stage of any conventional microscope. Although designed primarily for the analysis of petrological sections, it can be used for the analysis of any mixture, transparent or opaque, whose constituents can be recognized under the microscope. In effect the mixture is sampled at a fixed grid of points, the accuracy of the analysis being dependent upon the number of points sampled as explained previously under point counting.

With this device the selection of points is automatic, the operator having only to press recording buttons corresponding to the material observed at the intersections of a cross-lined reticle of the microscope ocular. As soon as the count is recorded, the stage is stepped electrically to the next point on the grid. A totalizing counter is provided so that uniform areas may be analyzed.

This type of accessory, as can be seen, is applied to controlling the positioning of a mechanical stage and at the same time tallying points counted.

2. <u>Differential count recorder for haematology</u>: This device, manufactured by Universal Optics Ltd. of England, applies an extended mechanical control to the fine focusing of a microscope. It has been designed to increase the speed and accuracy of blood cell counting.

The equipment is set up with a base plate on which the microscope is also placed (see Figure 98). The height of the connecting piece is adjusted, after loosening the set screw, so that a rubber ring slips over the fine focus control knob of the microscope. The microscope is aligned on the base plate at right angles to the connecting piece so that smoothness of fine focus control through the connecting piece is assured. The height adjusting set screw is then then tightened.

The knob on the connecting piece is now a continuation of the fine focus of the microscope. In addition, it serves as a counting knob so that focusing and counting may be accomplished simultaneously using only the forefinger and thumb of one hand, while the other hand manipulates the mechanical stage controls. By this means, the eyes and hands of the operator need not be removed from the microscope during the counting activity.

There are six possible positions of the selector/focusing knob which, on making contact, electrically actuate counters corresponding to the position selected.

While observing the specimen material and scanning the slide in a normal manner (focusing as required), when a relevant cell to be counted appears in the field of view, the selector knob is moved into a slot previously selected for that type of cell, as for example, Counter 1 slot 1 (a Basophil), Counter 2 slot 2 (a Eosinophil) etc.

After some practice, cells can be counted rapidly just as quickly as they are brought into the field of view. On reaching a total integrated count of 100, a buzzer sounds. By cancelling the totalizer button situated on the right-hand side of the bank of six counters, the buzzer is stopped. Each of the individual cell-type counts will be represented at its appropriate counter. These are

AUTOMATIC ACCESSORIES

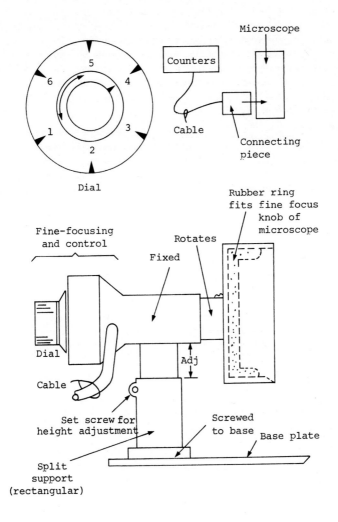

Figure 98. Differential count recorder.

read off and the counter cancelled to zero. Another complete operation can then begin with individual counts displayed and with a running total always present. The instrument can be adjusted for extending the total count to 500 rather than 100.

This and the preceding "automatic" counting apparatus described are representative of the accessory type that may be applied to the conventional microscope. There are others available from various manufacturers of a similar level of complexity for specialized applications.

D. AUTOMATED IMAGE ANALYSIS: Much of the characterization of specimen material with the microscope is dependent upon statistical analysis, and is very time consuming and error prone by manual methods. With the application of electronic methods, especially video and computer techniques, such characterizations can be performed in a fraction of the time required by manual methods and with tremendously increased accuracy. In fact, electronic methods now render previously impossible tasks in research, quality control and routine clinical assessment, both feasible and economic.

The equipment available, coupled with a light microscope, provides for automated (fully automatic in some cases) measurements of number, area, length, perimeter, intersections, form factor, transmittance, reflectance, absorption, fluorescence, optical density and combinatorial results of many factors simultaneously, complete with (in some cases) computer programming and printout facilities.

Typical applications of such sophisticated equipment ranges from the characterization of iron and steel, brass and other metals to the analysis of chromosomes. Almost any investigation of microscopic material that includes characterization by geometric means is especially benefited by the use of such equipment.

1. Basic system: Reference is made to Figure 99. In very simplified form, this block diagram shows the basic components of an automated image analysis system.

    a. Scanning: The optical image of the specimen produced by the microscope is scanned by the electron beam of a vidicon or plumbicon camera tube. The electronic scanning process converts the optical image to a proportionate series of sequential electrical values which are introduced into an electronic processing unit. The rate at which the scanning process is accomplished is dependent upon the design of the particular sys-

AUTOMATED IMAGE ANALYSIS

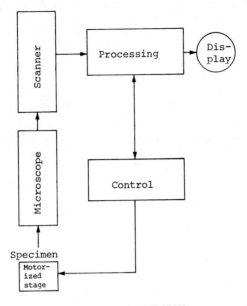

BASIC BLOCK DIAGRAM

Figure 99. Automated image analysis system.

tem. Conventional television scanning employs 525 lines (U.S.) and 625 lines (Europe), interlaced at 30 frames per second. This scheme, primarily designed for the scanning of moving pictures or images, offers flicker-free viewing and maximum interface compatibility with existing electronic systems.

Slow scan rates using 720 lines and 10.6 frames per second, noninterlaced, are especially adaptable to optimize precise and accurate image analysis of static fields of view. Both systems are available.

Almost any kind of microscope that can provide an image to be scanned is usable, although specific systems are usually designed around certain types or styles of instruments.

b. <u>Processing</u>: Electronic processing of the electrical equivalent of the optical image allows a great variety of analysis methods to be accomplished. The optical image, bit by bit, can be electronically coded, rearranged, compared, analyzed and displayed in many different ways. Digital control of processing circuitry is capable of, in some cases, dividing the image into as many as 650,000 discrete picture points which can be sampled individually. Measurements of size, size distribution, optical density etc. are accomplished at very high rates.

   c. Motorized stages and search patterns: The specimen is moved about under the microscope objective typically by a motorized stage. Small electric motors drive the stage in both the X and Y directions. The control of the stage-drive can be exercised through manual switches (pushbuttons) or even through programmed automated means.

 Some typical search patterns of such stages are shown in Figure 100. The patterns used are selected on the basis of material or specimen composition, and the results desired in the analysis.

   d. Control: The various tasks performed in the processing and display of specimen material are ordinarily under the direct control of the operator-microscopist. The methods of control are various, dependent upon the system and the requirements to fulfill.

   (1) Switch controls: This type of control is either by positive on-off action, locking in either or both modes, or may be of a momentary contact type, exercising supervisory control of lockup relays or solid-state switching circuitry. Usually such controls are reserved for system settings and setups, and/or initiation of functions.

   (2) Joy-stick: This is a control which usually allows the operator a continuous variable selection of display features, usually in the form of vertically oriented levers which are adjustable in a universal frame of movement to typically control the size of a variable frame outline and to select discrete sizes and shapes for counting, sizing or measurement.

   (3) Light-pen: An electronic device that is positioned by the hand of the operator to select specific image components for particular measurements. Figure 101 illustrates a light pen being used to select a particle for area measurement. In the B & L QMS, as the light pen is moved across the displayed field, an electronic flag (or tag) follows it. As this flag is brought to rest on a component, the component becomes bright-outlined for the particular measurement to be performed. The

AUTOMATED IMAGE ANALYSIS

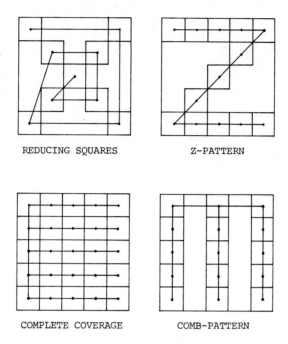

Figure 100. Motorized stage search schemes.

light pen is then pressed to record the result and the proper measurement figure will appear on the display and/or other registers as the equipment provides. More than one measurement can be made, if desired, on each feature. Types of measurements commonly made on individual particles are shown in Figure 102.

(4) <u>Automatic programming</u>: With the more sophisticated systems, completely automatic operations can be performed by interfacing computer equipment. All operations of an analysis can be automatically controlled including motorized stage movement in predetermined search patterns and including the number of steps to be made by the motor drive stage in both the X and Y directions. In addition to "line-scanning" and conventional search patterns, some systems provide for a "meander" type scanning program. Control is also exercised on predetermined dwell-times between scanning lines etc.

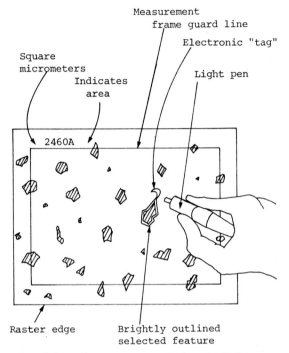

Figure 101. Selected measurement feature by light pen. (drawing adapted courtesy Bausch & Lomb.)

The processing equipment is likewise programmable for a variety of special or unique analysis tasks. Quite often, such equipment is initially designed to perform certain types of standardized analyses without special programming. In some systems, unique or special programming is introduced via teletypewriter or computer interfacing equipment, software or other means, with programs written either by the user or manufacturer. Figure 103 illustrates part of a computer readout on an analysis of particles in a steel sample.

2. Data results: Most of the systems coming under the heading of automated image analysis provide, in one form or another, the following results. The results generally desired (and provided) in this type of work are information per field and information on individual or selected particles.

a. Field measurements: The type of measurements

AUTOMATED IMAGE ANALYSIS

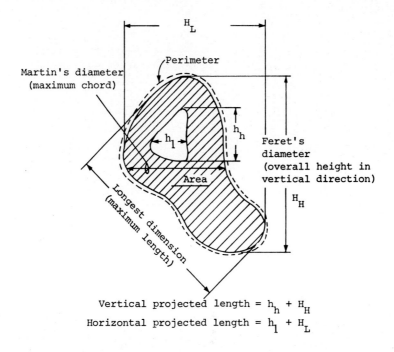

Figure 102. Some particle measurement terminology.

performed on a complete field of view might include, but are not necessarily limited:

    (1) total count.
    (2) total area.
    (3) average area.
    (4) percent of area.
    (5) total intercepts.
    (6) total number of tangents.
    (7) average projected length.
    (8) total projected length.
    (9) oversize counts (counting all particles over a given size).
    (10) counting all intercepts larger than a selected length.

    b. <u>Selective particle measurements</u>: Selective particle

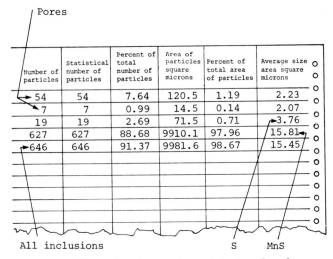

Figure 103. Example of computer printout analyzed particles in steel sample. (drawing adapted courtesy Lemont Scientific.)

measurements include but, again, are not necessarily limited to those illustrated in Figure 102. These are usually displayed on the monitor, being marked electronically, outlined appropriately and digitally indicated.

  c. <u>Field multiple measurements</u>: Field multiple measurments and statistical results are also possible and include:
   (1) count ratios.
   (2) feature counts.
   (3) grain size density distributions.
   (4) sorting distributions by sizes.
   (5) pattern recognition, separation and classification.
   (6) frequency distributions.
   (7) volume percentages.

  3. <u>Displays and presentations</u>: The conventional display for the automatic systems thus far discussed is the cathode ray tube or kinescope (Figure 101). The advantages of this type of display are many. Perhaps the outstanding advantage is that, by electronic means alone, the display can be manipulated to provide a variety of borders, outlines, indicators, digital annotation and

other visual aids to the operator. For permanence, photographic records of such displays can be made.

When computing equipment is made a part of these systems, in addition to the visual monitor display, the results of various analyses are presented in printed form via a teletypewriter output. The resulting data may be in digital form, in tables of size distributions or type-generated histograms on sheet paper or on punched tape. Communication with computing equipment components is also effected by the teletypewriter terminal.

Other forms of data recording are available with some systems including digital printers, magnetic tape recorders, pen recorders and desk-top calculators.

4. Discussion: It can be seen even by this very superficial discussion of automated image analysis that it is a very complex, very far reaching technique that is blooming into a field of its own. It already has become an agglomeration of a great number of other specializations, such as mathematics, physics and electronics, with a special jargon for communications of its ideas, concepts and operations. It is continually borrowing and adapting the most up-to-date and modern of our technologies. Its advance in the past ten years is enormous, and the future is certain to reveal even greater strides. It would be very difficult to master even one of the specialties involved in the field, let alone the gamut of the wide range of technologies involved. We can but hope to stay "informed". Other volumes of this series take up this fascinating subject in much more detail.

E. REFERENCES AND COMMENTARY

1. Stroud, A., J. Butler and M. Bütler, "The combination of microscopes and computers for the analysis of chromosomes," The Microscope 15, Third Quarter (July 1967).

2. Snedecor, George W., Statistical Methods, Collegiate Press, Inc., Ames, Iowa, 1937.

3. Bayard, M., "Microprobe analysis of small particles," The Microscope 15, Third Quarter (July 1967).

4. Fisher, C. and M. Cole, "The metals research image analyzing computer," The Microscope 16, Second Quarter (April 1968).

5. Arkin and Colton, Statistical Methods, College Outline

Series, Barnes and Noble, Fifth Ed., 1970.

6. Fisher, C. and L. J. Nazareth, "Classified treatments for the application of the Quantimet to stereological problems," The Microscope 16, Second Quarter (April 1968).

7. Humphries, D. W., "Particle size measurement and a new semi-automatic recording eyepiece micrometer," The Microscope and Crystal Front 15, No. 7 (November-December 1966).

8. Cole, M., "The metals research Quantimet (QTM)," The Microscope and Crystal Front 15, No. 4 (May-June 1966).

9. Widdowson, R., "Automatic counting of non-metallic inclusions in steel," The Microscope and Crystal Front 15, No. 3 (March-April 1966).

10. Neuer Harald, Quantitative Analysis with the Microscope, Zeiss Information No. 60, 1966.
 A good basic treatment of point counting.

11. Lark, P. D., "Semi-automatic particle counter," The Microscope and Crystal Front 15, No. 1 (November-December 1965).

12. Watts, J. T., "Micro-stereology," The Microscope 18, First Quarter (1970).

13. Brown, J. F. C., "Automatic microscopic analysis with the πMC particle measurement computer," The Microscope 19, Third Quarter (1971).

14. Jesse, A. "Quantitative image analysis in microscopy — a review," The Microscope 19, First Quarter (1971).
 Very good article. This entire volume is devoted to the general subject.

15. Grosskopf, Rudolf, Microscope Photometry and Stereometry with the Process Computer, Zeiss Information No. 80, 1972/73.

CHAPTER 6

MICROSCOPY AS ADJUNCT TO OTHER TECHNIQUES

A. <u>INTRODUCTION</u>: In this chapter will be included a selected group of special and supplemental applications of the light microscope. Light microscopy is ordinarily concerned with the obtaining of information from very small structure and detail. The conclusions drawn from observations of that type are applied in identification, analysis and in general increase our knowledge of the animate and inanimate world. Classic activity can be considered as either an applied or research use of the instrument as the primary device of investigation or accomplishment.

There are other areas in which the light microscope either plays an extremely specialized role, or in which it assumes a supplementary or attending position in the overall scheme of activity. In these areas, it may assume very radical or innovative forms, or be integrated into instruments which have purposes far removed from light microscopy as such. The light microscope assumes a supporting or supplementary role in those areas that, in most cases, is not unimportant. In fact, it is ordinarily an essential element in the prosecution of the activity, or one which would be sorely missed if absent.

We can also include its use as a full partner in many areas of microscopical investigation. For instance, in electron microscopy, electron microprobe, ion microprobe and similar techniques of specimen characterization and examination, the analysis and final conclusions regarding the characterization are often not complete without thorough optical light microscope investigation as well. Some of these techniques wherein the optical light microscope is often fully involved are also included.

It is these adjunctal uses of the light microscope that are included herein to give the uninitiated, or even the average working microscopist, appreciation of the far-reaching impact of this king of scientific instruments. It is not possible, of course, to include in this limited space all of the uses to which the light microscope has been put. However, there is included a selection of the more important (and potentially important) ones, and a sufficient diversity of applications as to indicate scope.

B. <u>INDUSTRIAL OPERATIONS</u>: Industrial operations that require microscopes in various forms as adjuncts are those mainly associated with manufacturing. Fabrication, assembly, quality control and testing all utilize light microscopes in various specialized forms and/or procedures for their purposes. In the following paragraphs we will cover some of these applications of the light microscope as important, even indispensible contributors, to the successful prosecution of industrial activities. In the discussion we are omitting the industrial research uses of the instrument which fall mainly in the area of scientific or applied scientific research, whose techniques and instrumentation are common with the sciences. Concentration will be on the instrument in special and specific application as an auxiliary to other than pure microscopical observation, interpretation etc.

1. <u>Measuring microscopes</u>: Measurements connected with manufacturing operations include linear and angular measure much the same as in the biological and other sciences. However, because of the diverse size and shapes of manufactured items, there are some special problems, and special methods and apparatus to solve them.

In the sciences, the metric system has been applied universally, and measuring accessories and mechanical stages, locating devices etc. are commonly graduated in metric intervals. In the United States and Great Britain, adherence to the English (FPS) system in industrial activities requires microscopes with corresponding calibrations. In European countries and other parts of the world, the metric system prevails and measuring microscopes are graduated and/or calibrated in that system. In our brief discussion here, whether we refer to English or metric units, it is understood that the equivalent range and accuracy can be obtained in both systems.

The continual narrowing of production tolerances in the metal-working industry has brought with it a need for very precise measurements. Microscopes, separately or as a part of larger apparatus, are ideally suited to making measurements and positioning adjustments to a much finer degree than by any other means.

Measuring microscopes (Figure 104) as used in industry are usually calibrated in the English system in the U.S. and are of two general types:

       (1) A moving microscope with a stationary stage.
       (2) A stationary microscope with a moving stage.

The means of measurement may be of two kinds:

       (1) A scale micrometer microscope with the object measured by a scale.
       (2) A microscope equipped with a filar micrometer ocular.

INDUSTRIAL OPERATIONS

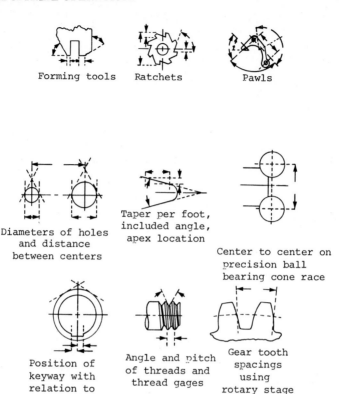

Figure 104. Some types of measurements by measuring or toolmaker's microscope. (adapted illustration courtesy Gaertner Scientific Corporation.)

The advantages of using microscopes as measuring, positioning and observation devices in industry are much the same as in the sciences. For instance, a turning or milling operation may be observed, measured and/or positioned without contacting the material, even though the dimensions etc. are in a continuous state of change. In addition, for very large surfaces, the microscope can be moved about in accordance with the work or process conditions, sequentially or continuously.

The precision of measurement generally will be limited from a fraction of a ten-thousandth of an inch to a fraction of a hundred-thousandth of an inch. A typical measuring microscope is of comparatively low magnification (30X to 35X) and can be reset to within about 0.00005 inch.

Because of the need for large working distances, a minimum illumination requirement and maximum insensitivity to disturbances, microscopes used in measurement work in manufacturing (fabrication especially) have assumed some special characteristics in design and application that are noteworthy.

Typically, the objective-ocular magnification is the lowest possible that can be used, even if it is necessary to obtain the last desired figure of a reading by interpolation. This procedure assures the most favorable of the conditions above. Objectives are usually achromatic doublets and oculars of the Ramsden or orthoscopic type, the latter being better corrected for flatness of field (Figure 105).

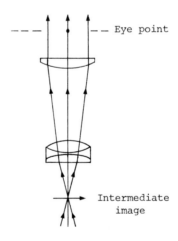

Figure 105. Orthoscopic ocular.

Accessibility of the object under measurement and the type of measurement are the determinants as to what type of optical-mechanical arrangement will be most desirable, as the microscope in many cases is taken to the work or oriented so as to ex-

INDUSTRIAL OPERATIONS

amine the work during fabrication. In other words, the microscope is adjusted or suited in design and orientation to the requirements of industrial operations, whereas in most microscopical work the object is prepared and adjusted to suit the microscope conditions.

The schematic of Figure 106 indicates the arrangement used

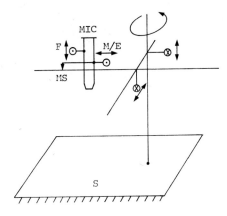

O	Continuous control
⊙	Interval control (locking)
F	Focus
M	Measure
E	Examine
S	Worksurface (stage)
MIC	Microscope
MS	Micrometer slide (scale)

Figure 106. Moving microscope. Fixed stage.

in a moving-microscope, fixed-stage setup. Although shown for examining and measuring horizontal objects, it also can be applied to work in a vertical plane. Interval controls of the locking type allow the microscope to be adjusted in height and in a bilinear direction. If the support is a round pillar, a rotational interval adjustment is also possible.

The microscope is supported on a carriage whose lateral movement is controlled continuously by a handwheel or knob-driven screw. The ways upon which the carriage travels are graduated as is a drum coaxial with the screw. The range over which such measurements can be made is from 1 to 8 inches.

Precision with such an instrument can be as high as 0.00005 inch with a vernier scale. Illumination necessary for reading the scales either has to cover the distance traversed, or the microscope itself must carry an illuminator. A very important consideration is that the ways upon which the carriage travels must be accurately straight. Any tilt will result in large measurement errors.

The carriage assembly on which the microscope proper mounts in such "travelling microscopes" is more commonly termed a micrometer slide. The design and construction of this mechanism is specifically directed to accuracy, smooth uniform motion and long service. To this end, specially selected tool steel for the slides, screws carefully ground, lapped, checked and corrected in every case, the use of nuts lapped to the screws they are used with, and ingenious mechanical design to correct residual errors of the screws are incorporated in this type of carriage. Each precision screw is calibrated by comparison with a standard scale, the accuracy of which may have been certified by the National Bureau of Standards, Washington, D.C., or the Bureau International des'Poids et Measures, Sevres, France. Certification of calibration is ordinarily furnished with such instruments.

2. The cathetometer: The cathetometer is a special type of moving microscope used in the accurate measurement of vertical distances. This instrument usually uses a scale instead of a measuring screw.

Because of the distance between the cathetometer and the object, very slight irregularities in the ways can cause errors. Therefore, a sensitive striding level is placed on the microscope to detect tilt. The scale used for measurement is either on the vertical column or placed at the same distance as the object under measurement. In this latter case, the observation alternately of the scale and the object is affected by rotational capability of the vertical guide column. Some configurations employ a tripod for floor mounting. When a very distant or inaccessible object is to be measured, the microscope is often replaced by a telescope.

3. Toolmaker's microscope: One of the most elaborate and

INDUSTRIAL OPERATIONS

effective applications of optical methods in technical measurements is the toolmaker's microscope (Figure 107). It is essentially a fixed microscope, moving-stage style of instrument. Length can be measured in two directions in the horizontal plane and angle in any direction. This capability enables complete measurements to be made on jigs, tools, screws, hobs, tapers and many other parts difficult, and sometimes impossible, to gauge by any other method.

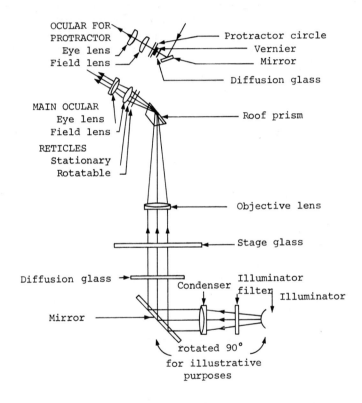

Figure 107. Optical system.
Toolmaker's microscope.
(illustration courtesy Gaertner Scientific Corporation.)

The greatest difference between this type of instrument and similar types used in scientific pursuits is the elaborate stage construction. The entire instrument is built in massive proportions, and the stage is proportionately even more massive and complex in its design. The stage typically moves on compact slides which travel on precision ball bearings preloaded in hardened and lapped steel guides. This type of design provides for long life wearing properties under constant use and heavy loads. The lower slide moves longitudinally on the base while the upper slide provides the cross motion for coordinate measurements. Each slide is operated by a micrometer screw, the floating nut having a hardened steel insert maintained in contact with its respective slide by spring tension. These movements provide for a linear measurement when the object is placed on the stage.

TABLE XI

Specifications
Typical measuring microscope optics

Magnification	Equivalent Focal Length (mm)	Diameter of Field (mm)
RAMSDEN EYEPIECES		
20X	12	10
14X	18	10
10X	25	11
6.7X	38	14
5X	50	14
ORTHOSCOPIC EYEPIECES		
14X	18	10
10X	25	12
WIDE FIELD EYEPIECE		
10X	25	22

MICROSCOPE OBJECTIVES

Equivalent Focal Length (mm)	Initial Magnification	Optical Tube Length (mm)	Mechanical Working Distance (mm)
8	21X	177	1.8
16	9.6X	170	7.5
25	5.1X	153	17.
32	4.0X	160	35.
38	3.2X	160	45.
48	2.3X	160	65.
60	1.7X	160	89.
67	1.4X	160	109.
80	1X	160	155.

(information courtesy Gaertner Scientific Corporation.)

INDUSTRIAL OPERATIONS

A typical range of movement for such an instrument is 4 inches of longitudinal motion (X) and 2 inches of cross motion (Y). The micrometer drums are commonly graduated to read 0.0001 inch per division. The stage can usually be rotated about 5° by means of a tangent screw in micrometer form to read rotation in tenths of a degree.

Some manufacturers provide integral or rotatable stages having a full 360° range. To measure angles which lie within the field of view, the stage is centered by the X-Y controls and the apex of the angle on the workpiece is placed at the center of the crosslines in the ocular. The angle generated by the stage is read then from an engraved scale on the stage itself.

a. Measuring reticles: When difficulties with size and orientation of odd-shaped object materials are encountered, advantage is taken of the use of reticles in the ocular of the microscope. The reticle of the toolmaker's microscope ordinarily consists of two thin glass plates, individually rotatable, the engraved sides of which face one another with very little separation. One plate carries a zero line and a short protractor scale, and is rotated by a knurled ring around the ocular. The other plate is engraved with a single line and turns with a large protractor circle which is fully enclosed in a large flanged housing around the ocular. This arrangement facilitates accurate point-to-point settings on angular objects (such as screw-thread profiles).

The ocular is inclined at a convenient angle facilitated by a prism which also produces an erect and correct right-left image. The microscope support column can be tilted for accurate settings on the contours of screw threads (Figure 108).

The reticles used in the measuring and toolmaker's microscopes are not restricted to the ones previously described. Figure 109 illustrates some of the different types to suit various applications of the instrument. The most versatile of the illustrated fixed patterns is that of A. Settings are made on wide lines by balancing visually the small triangular portions of the line which appear left and right of the crossline junction. If the line (as it appears in the field of view) is the same weight (width) as the crosslines, then setting is made on the basis of bisecting the triangular spaces above and below the intersection. This configuration takes advantage of the fact that it is easier for the human eye to judge accurately the relative differences in patterns of spaces than the coincidence of two points or lines.

For special applications, the reticle pattern at B (parallel

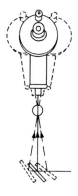

COLUMN TILT

Figure 108. Toolmaker's microscope. (illustration courtesy Gaertner Scientific Corporation.)

crosslines) may be even more accurate than the 60° pattern at A. This obtains when the magnification of the microscope provides an image of the line to be gauged that is about half the width of the spacing of the parallel crosslines (Figure 110). Again, the human eye is very adept at judging the balance of the very narrow spaces on each side when the object line is centered. If the magnification is too great, the width of the object line will blank the crosslines. If too little, the spaces each side of the object line will be too great to judge as accurately.

For measuring purposes the 90° reticle pattern at C is not as useful as the previous ones. In some cases, however, the horizontal line assists in alignment with a similar line on the object. A simple crossline is of some value in setting on a bright line in a darkfield.

The crosslines of the toolmaker's microscope are usually rotatable with respect to one another through 360°. The rotation is measured in one-half degree intervals by the reticle protractor scale, and to 1 minute of arc by the large protractor scale on a flanged housing external to the ocular. Settings can be made in the same manner as with the 60° fixed crossline reticle, or the crosslines can be adjusted to the contour of the object outline as in D in Figure 109.

With a filar micrometer ocular, a measuring microscope

INDUSTRIAL OPERATIONS 283

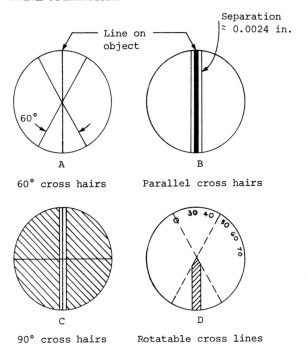

Figure 109. Reticle types. Toolmaker's microscope. (illustration courtesy Gaertner Scientific Corporation.)

with reasonable care can be set to 0.00001 inch and no serious difficulty should be experienced in measuring objects as small as 0.0001 inch in size. This type of ocular is very similar to those employed in conventional research microscopes excepting for the English measurement intervals (U.S.) and, in some cases, there is in the field of view a "comb" or revolution counter. Special assemblies can be obtained with an external revolution counter of the dial type. The type of dial scale pattern is governed, of course, by the scale-type on the micrometer drum.

A typical field of view through such an arrangement is shown in Figure 111. Great precision of measurement can be obtained by using this style of instrument with a standard scale, the microscope serving to divide the scale.

In measuring microscopes, a protractor-ocular is commonly available to measure angles. This accessory has a circle divided

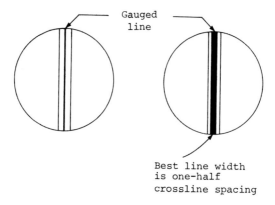

Figure 110. Effect of magnification on setting with parallel crosslines.

in degrees. When rotated, the circle carries with it 90° crosslines with which settings are made. The angular position of the circle is read by a vernier to 3 minutes of arc. Fine rotation adjustment is ordinarily effected by a small pinion which engages a toothed edge of the circle. Linear and angular measurements can be made simultaneously by mounting a scale or filar micrometer on the protractor-ocular.

Calibration of the reticles of the toolmaker's microscope and other similar industrial instruments is accomplished by a stage micrometer or calibrated scale. Typical of these is a stage micrometer scale 10 mm long with 100 divisions, each of which is 0.1 mm long with every fifth and tenth line extended. This scale is mounted lengthwise in the center of a 25 x 75 mm glass microslide. A similarly marked scale can be obtained mounted crosswise near one end of a 25 x 75 mm microslide, making it more convenient in such applications as on the carriage of an interferometer. Similarly mounted scales in English units are also available. A 0.4 inch scale divided by 100 divisions of 0.004 inch in length is a common design.

For calibrating the screws of the micrometer stage of a toolmaker's microscope, standard scales on glass are used. The scale is etched on a glass plate with a liberal free length at each end of the ruling. A standard metric scale typically is 50 mm in length with divisions of 0.5 mm, every centimeter being num-

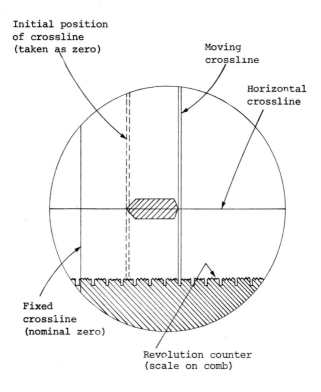

Figure 111. Field of filar micrometer ocular.
(illustration adapted courtesy
Gaertner Scientific Corporation.)

bered. Scales are certified by the manufacturer and can usually be certified back to the National Bureau of Standards.

    b. <u>Lighting</u>: Lighting requirements for a toolmaker's or measuring microscope may be quite variable. Vertical incident illumination for shadow-free lighting and high-power examination of opaque objects, substage illumination providing collimated light for viewing contours and transparent specimens, and oblique spot-illumination for surface lighting of relief structures are all requirements in such application. For enhancement of contrast it is also valuable to employ a second oblique illuminator to accomplish a "duo-color" lighting of scribe marks and reference holes (Figure 112). In measurement work monochromatic lighting promotes greater accuracy and, therefore, suitable filters are sometimes a useful accessory to the lighting system. Shadowless dispersed illumination for certain surfaces is desirable. Ring-type illuminators, fitting around the objectives, are very useful in this respect. For projection and photomicrography, high intensity lighting is required. Some manufacturers of this style of instrument provide accessories for those purposes.

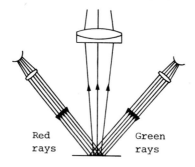

Figure 112. Bi-chromatic surface illumination. (illustration courtesy Gaertner Scientific Corporation.)

c. <u>Accessories</u>: Special accessories for the toolmaker's microscope are available to facilitate the application of that style of instrument. Inclinable stages and workholders are provided with adjustable spindles to accommodate screws, taps, hobs, millers and other cutting tools, as well as a glass stage plate to hold racks and other types of gears. In use, the cradle is inclined to the helix angle to provide a sharp contour which can be measured or compared with a standard pattern. Cradles available are inclinable ±15° and are calibrated in degrees.

Threads may be checked for form, angle, depth, lead and diameter, using a micrometer and rotary stage or, more conveniently, comparing the thread outline directly with a master pattern reticle in the ocular. By means of stage motion controls, the tool is moved to thread coincidence with any of the patterns. Special turret Screw Thread Oculars are available where frequent checks of this type are required. In use a knurled ring is used to bring into the field of view a succession of thread patterns for comparison and matching. In addition this special type of ocular often has a scale to measure the angular deviation of the drunken threads.

Special stage inserts are available such as clamping jaws, center holder, collet chuck etc. for holding or orienting small parts for rapid examination and measurement. These are of special use in the watch and small components industries. Other accessories include tubular and circular levels of the bubble type with specified sensitivities as to the angle of tilt required to move the bubble per division. With tubular levels, sensitivities of 10 seconds of tilt with repeatability to 5 seconds are available.

When measuring the distance between an edge and some other part of an object, it is difficult to set a microscope accurately on such an edge bounded by a plane. To facilitate such settings, surface index blocks can be obtained as accessories (Figure 113). The blocks have an index line ruled on top in the plane of the inner surface which is brought against that edge of the work from which the measurement is to be made.

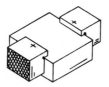

Figure 113. Surface index blocks. (drawing adapted courtesy Gaertner Scientific Corporation.)

d. <u>Digital readout</u>: An electronic digital readout attachment is available for Gaertner Scientific Corporation microscopes. It consists of two encoders, and an electronic digital counter and readout can be provided in either English or metric units. Direct readout is to the nearest micrometer or to the nearest 0.0001 inch.

The encoders are attached to the precision micrometer lead screws of the microscope via adapters and special handwheels. Cable connection is made to the digital counter-display unit. The counters are bidirectional allowing the shaft to rotate in both directions, thus permitting back tracking. The display can be reset to zero or preset to any 5 digit number at any point by means of thumbwheel switches. This allows the operator to choose a zero point, selecting any desired position and then taking readings from that point. After selection of the zero point readings above that point will be indicated by a plus sign, while readings below that point will be indicated by a minus sign. A reversal switch is provided for each axis to permit reversing the direction of reading on the counter while maintaining the same direction of rotation of the micrometer screw. The counters will accept pulses at rates up to 20 kHz.

In addition, the electronic counter includes a set of mating connectors to permit the use of auxiliary recording equipment such as a printer, paper tape, punched cards or magnetic tape. The output is Standard Binary Coded Decimal (BCD). Figure 114 illustrates this elaborate instrument.

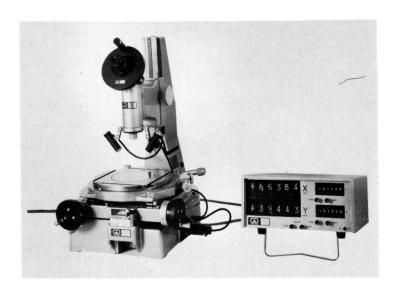

Figure 114. Toolmaker's microscope. Illustration shows inch units on the x-axis and metric units on the y-axis. (photo courtesy Gaertner Scientific Corporation.)

4. <u>Depth measurements and surface finish</u>: Microscopic depth measurement is much more common in industrial activity than in the sciences. Requirements for this type of measurement include, but are not necessarily limited to, depth of pits, grooves and scratches, wearing patterns and milled features, as well as depth of undercutting in chemical blanking, and depth of etch in silicon transistor wafer construction and similar fabrication techniques. The printing and graphic arts industries also have need for critical examination and quality control of etching wells in Ballard shells, intaglio cylinders, and all kinds of printing plates.

The surface finish microscope is used to examine surfaces for degree of roughness of finished heavy duty parts and semi-finished products, as well as measuring depressions, grooves, slots etc. on items purposely so worked. The degree of tolerance to which these types of examinations/measurements must be made require magnifications from 50 to as high as 500-600 diameters. Monocular or binocular widefield oculars, providing as much as a 50% larger field of view, are used with this type of instrument.

In one style of this instrument marketed by Unitron Instrument Company of Massachusetts, the depth measurement is made by taking the difference of two readings from a depth gauge. The gauge is of the dial type coupled to the fine focus control of the microscope, reading directly in units of 0.0001 inch. The accuracy of such measurements is increased at high magnifications wherein the depth of field of the objective lenses are very shallow, providing very definite "decision levels" of focusing.

Zeiss markets similar equipment. Common to their many models is a large drum-type fine focusing control with screw-micrometer type markings. Difference of settings and readings of this control determine the depth measurement. Additionally, a coupling control allows a zero-set at the high elevation of the object area to be measured. Then, when focused on a respective lower point, the reading from the micrometer drum scale is the height or depth measurement without any further computation. This feature is accurate to within 1 μm with estimations possible to a fraction of a micrometer.

A depth measuring or surface finish microscope is equipped variously with stands that allow it to be mounted directly on very large workpieces as well as for the examination of small objects. By means of rollers the microscope may be mounted on machine shafts for their inspection or on other turned or tubular parts or

Figure 115. Toolmaker's microscope with depth gauge and stage cradle. (photo courtesy Unitron Scientific Inc.)

curved surfaces. Zeiss produces a ball-stage that allows inspection from different directions, the instrument being tilted in a supporting ring. Column stands similar to those used with other types of measuring microscopes are also found to be useful and allow 360 degree rotation. Some stands provide a lateral (X) or (Y) adjustment of the microscope proper relative to the stand in increments as small as 0.01 mm. This facilitates relocation of previously observed features.

Lighting requirements in most cases involve vertical or oblique illuminators, depending upon the surface features to be examined or measured. The intensity, controls and other features of such lighting facilities are in every respect the equal of those used in the sciences, including provisions for multiple-

INDUSTRIAL OPERATIONS

beam interferometry.

For reference and for legal records of quality control or adherence to contract specifications, these microscopes usually have provision for 35 mm, or Polaroid camera accessories for taking photomicrographs.

As an example of the ultimate in specialization of surface inspection instruments, consider the "Hole-Inspection" microscope

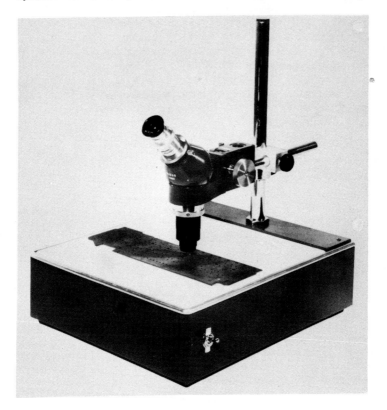

Figure 116. Hole-inspection microscope.
(photo courtesy Nikon.)

made by Nikon. This unique instrument is equipped with special optics for inspecting wall surfaces of holes 0.002-0.375 inch in diameter and up to six inches in depth. A zoom feature shows a sharp 360° image of the cavity. The microscope is mounted on a pillar and the stage or base is from 6 x 8-1/2 inches to 10 x 10 inches with a translucent glass surface illuminated from below with fluorescent lamps. A special carriage style of mount can be provided that allows extensive scanning of holes in workpieces as large as 10 x 23 inches in dimension.

This specialized instrument will immediately reveal "bleeding" between multilaminate materials pierced with holes. For blind-hole inspection a vertical illuminator is available. The startling effect of viewing holes with this microscope is that of driving into a brilliantly lit tunnel. Focusing takes you farther and farther into the hole, increasing the image size to fill the entire field of view while maintaining sharp focus throughout.

A similar purpose instrument has been developed by Reichert of Vienna, Austria. The optical system consists of a spherical reflecting surface below the bore-hole which produces an intermediate image of the bore-wall and a microscope focused on this intermediate image. Figure 117 illustrates the layout of one

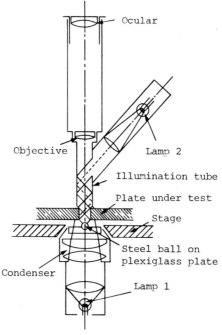

Figure 117. Essentials of a hole-inspection microscope. (illustration courtesy Microscope Publications.)

version of such an instrument. The upper lamp furnishes illumination for the upper adjacent surface and bore-wall, and the lower lamp illuminates the underside adjacent surface to the hole through the transparent support surface. Standard incident-light brightfield/darkfield microscopes can be modified for the purpose, utilizing the basic principle of the "fish-eye" objective.

5. Comparators: When measurements are to be made over a large range, especially when coordinate measurements are involved, it is more convenient for the observer if the microscope remains fixed while the object is moved via a mechanical stage, much the same as a toolmaker's microscope. Specialized instruments have been developed, each adapted to a specific class of measurements.

For example, large stage travel is required in examining photographic films and plates for very precise determination of position, change of position, or paths of astronomical bodies, or in geological measurements as applied to photogrammetry.

Just as the toolmaker's microscope is basically a coordinate comparator with certain additional features for toolroom inspection and precision machine shop measurements in general, the linear comparator is commonly used for measurements on spectrographic or x-ray diffraction plates.

Linear comparators are typified by very large sloping surfaced, moving stages of massive construction with equally massive controls, much the same as are found on extremely accurate machine tools. Massiveness of construction imparts rigidity and mechanical strength qualities which protect against vibration, thermal and setting expansion, and other physical phenomena adverse to accurate and repeatable measurement.

A single-axis digital readout system that mounts on the lead screw of Gaertner micrometer slides can be adapted to most of their precision linear comparators. The convenience, flexibility and accuracy of such an arrangement described for the toolmaker's microscope is thus extended to these instruments as well.

6. Extensometers: For measuring dimensional changes of specimen materials that are due to creep, thermal expansion, strain and other environmental conditions, the measuring microscope is well suited. When especially equipped and mounted it is referred to as an extensometer. The minimum special requirements for this application are that the measuring elements be absolutely rigid, that they do not affect the specimen and that they

undergo no dimensional changes even during extended time-interval tests. Specimens are often in relatively inaccessible locations, including furnaces, and special "relay" lenses and illumination may be required. The various types of supports necessary are dictated by the task to be accomplished. Illumination of the specimen, while in a furnace for instance, may be a vertical or on-axis type, supplied through the microscope objective, if the subject material is a flat polished surface. If the material is notched and/or otherwise susceptible to background illumination, sometimes the furnace itself is provided with a back window for illumination purposes. For cylindrical samples and specimens not suited to background (brightfield) or vertical illumination, some type of off-axis illumination is required.

During certain types of tests, the specimen materials themselves may become self-luminous due to high temperatures, perhaps even excessively so. In that case, Polaroid filters at the ocular are used to reduce the light intensity at the eye of the observer.

C. MEDICAL PRACTICE: In various phases of medical practice, as distinguished from medical research, the microscope serves as a valuable adjunct, and in a number of cases assumes special forms with special attributes and apparatus associations. In diagnostic, observational and surgical techniques, the microscope has become an indispensible aid to the physician.

1. Operation microscope — surgery: Perhaps the most well known of microscope applications is in the area of micro-surgery of the ear, eye, nose and throat. Stereomicroscopes are almost invariably used in this application as the inherent three-dimensional upright and unreversed image of such instruments is of major importance.

Special requirements for stereomicroscopes utilized in this field are both optical and mechanical. Extremely long working distances desired require special objectives of focal lengths as great as 500 mm, providing a working distance of 470 mm (approximately 18.5 inches). Total magnification of more than 60X is possible at this working distance with proper ocular/objective selection.

Because, in many cases, microscopic observation must be effective within deep cavities and recesses of the body, special axial or semi-axial illumination of high intensity is required.

These instruments include special mechanical supports and controls so that they at all times are adaptable to varying physical situations with minimum effort on the part of the physician/observer. Most are supported on elaborate floor stands adjustable in height, and with universal movements at the selected height. Height adjustments vary, but usually range from 48-68 inches from floor level to the objectives. The microscope may be positioned from 6 to as much as 24 inches from the support column. Adjustments to height are generally counterbalanced to reduce effort, and illumination is often controlled by a foot pedal.

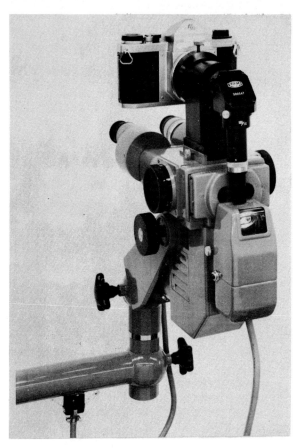

Figure 118. Operation microscope with camera attachment. (photo courtesy Olympus Corporation of America.)

Various viewing heads are available from a straight-away arrangement to 45° angled ones for greater viewing comfort. Provision is often made for camera attachment. In some systems, available separate viewing provisions are made for assistant surgeons or other observers to a particular operation. Zeiss has developed such instruments complete with motorized head for vertical movement controllable by foot-pedal switches. Because of the great utility and growing importance of this type of microscope to surgical practice, it is under a continuing state of development.

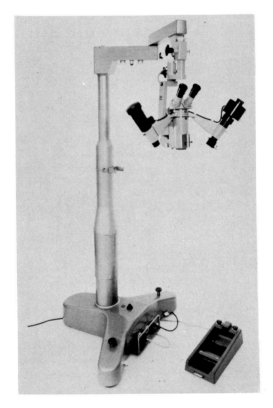

Figure 119. Floor mounted operation microscope for ophthalmology with foot switch elevation control. (photo courtesy Carl Zeiss Inc.)

MEDICAL PRACTICE   297

   2. Ophthalmology

       a. Fundus camera: In diagnostic work and treatment of
the human eye, the light microscope has become an indispensible
tool in the hands of the physician. In fundus observation and photography a specialized microscope with camera attachments is
termed a fundus camera. By guiding the patient's eye with a fixation device (a target such as a black point on the end of a fine

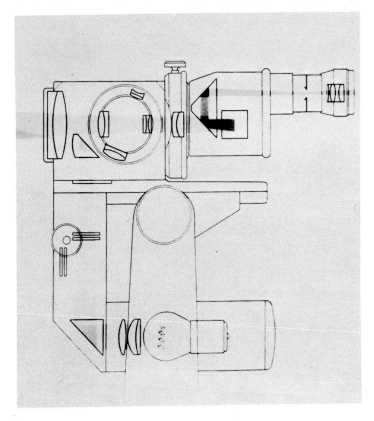

Figure 120. Illumination and viewing beams in operation
           microscope. (photo courtesy Carl Zeiss
           Inc.)

needle) that remains in view during observation and photography, it is possible to see and photograph various areas of the fundus. Also, the location and constancy of the fixation center will provide clues as to the nature of the visual defect and possible success of treatment, as the fixation point will be reproduced on the retina and photographable on film. Magnifications usually are in the range from 2-40X.

    b. <u>Slit-lamp</u>: Zeiss has pioneered in the development of slit-lamp photography of the eye. This instrument produces a

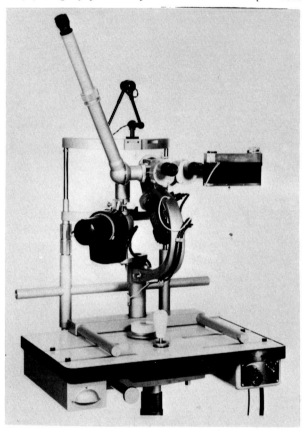

Figure 121. Photo-slit-lamp with beam splitter, 35 mm camera and observation tube. (photo courtesy Carl Zeiss Inc.)

MEDICAL PRACTICE   299

light slit of adjustable width in the anterior eye media (cornea, aqueous, lens and anterior vitreous). The light section produced can be observed from a certain angle with a corneal microscope (one similar to that used in fundus examination) and photographed for record purposes. Since this specialized method reveals conditions which cannot be detected under conventional illumination and the unassisted eye, it is an important diagnostic tool for the ophthalmologist.

During slit-lamp examination, the observer looks only at that part of the image in which the slit is sharply defined, and which therefore produces a fine section. By placing the slit image successively in various layers and areas of the media of the eye, he can select areas to photograph for record or study purposes.

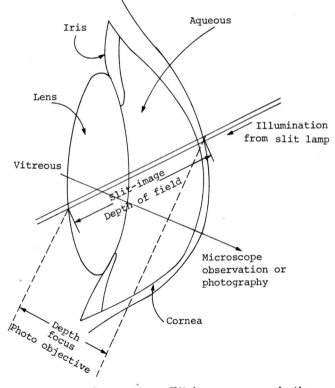

Figure 122. Crossection. Slit-image eye examination. (adapted drawing courtesy Carl Zeiss Inc.)

Difficulties with image contrast and obtaining sufficient depth of field of the slit-image are the main difficulties in such work. It is desirable to have a great depth of field of the slit image in order to examine as many layers as possible (with focusing of the microscope) and a very bright image so as to render very weak opacities visible.

The depth of field of a slit-image is the area within which the pencil of light does not exceed a certain arbitrarily determined width. The "permissible" width is assumed to be three times as great as the slit-image width. A slit width from 0.1 to 0.2 mm is used.

The apparatus in physical appearance is quite similar to that used in fundus work as the supporting mechanism is designed to accommodate examination of the same body element, the eye.

A schematic of the Zeiss photo slit-lamp appears in Figure 123. The width of the slit is adjusted by means of a lever and is usually at 0.1 mm for color photographs. The slit height is adjustable in steps of 9.5, 6.5 and 4 mm. All focusing and visual work is done using the tungsten lamp which is focused into the plane of the flash lamp by a double condensing system of lenses. During photography, the flash lamp supplies the necessarily more intense illumination.

The lamp, microscope and camera are so coordinated that their image planes have a common axis of rotation. The slit-image always lies within the object plane of the microscope and the object plane focused with the microscope will be imaged on the film plane of the camera.

D. X-RAY MICROSCOPY: The radiation from the target of an x-ray tube, when investigated by reflection at all angles from a crystal, is found to consist of a continuous spectrum and a superposed bright-line spectrum extending over a range of wavelengths from about 10^{-6} to 10^{-9} cm. The bright-line spectra are characteristic of target materials, and the spectrum for any one element consists of several well defined groups of lines. These groups are conventionally called the K, L, M, N . . . series, but the lighter elements give only the first mentioned ones. The spectra of all of the elements are similar, but the corresponding lines of any series occur at different wavelengths for the different elements. For instance, the wavelength of x-radiation for aluminum (Al) is K (8.3 Å). Reference is made to Figure 124 where the elements Cu, Mo, Sn and Au and their respective atomic numbers are shown with their characteristic spectra. Other L series

X-RAY MICROSCOPY

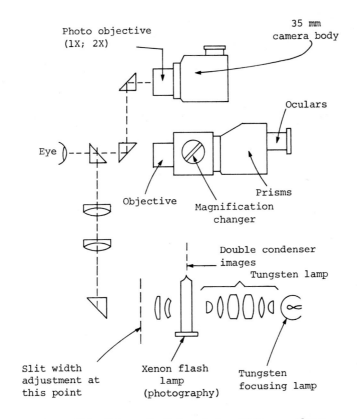

Figure 123. Schematic Zeiss photo-slit lamp. (drawing adapted courtesy Carl Zeiss Inc.)

spectra and beyond fall past the limits of the chart.

X-ray photons are not deflected in electric and magnetic fields and since the refractive index for them is so close to unity for any material, lenses in the conventional sense for focusing x-rays do not exist.

1. <u>Contact microradiography</u>: Placing specimen material in direct contact with recording material, exposing it to x-radiation and viewing the resultant image with a microscope is called con-

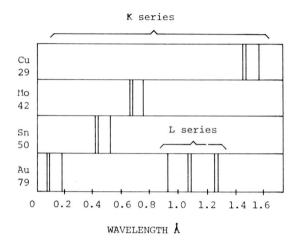

Figure 124. X-ray spectra — copper, molybdenum, tin and gold.

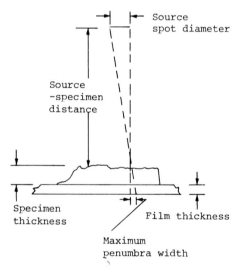

Figure 125. Some important geometric relationships in contact microradiography.

X-RAY MICROSCOPY

tact microradiography. Here, the light microscope is a viewing device of an image converted from one form of radiation to another. The x-rays produce changes in photographic film, for instance, in accordance with specimen characteristics. Then the film is viewed by conventional light microscopy for study and interpretation.

Fluorescent screens upon which the specimen material is placed are also used. In this contact method, the action of the x-rays on a fine-grained phosphor coating causes it to fluoresce in accordance with specimen characteristics. The advantage, of course, is that of being able to view directly with the microscope the image so formed. It also can be photographed for record purposes. Again, this is a conversion method wherein x-rays produce visible light by fluorescence that is focused and imaged by the light microscope (Figure 126).

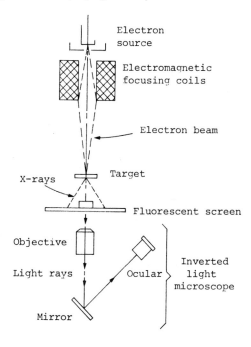

Figure 126. Essentials — schematic.
X-ray microfluoroscope.

Other materials such as plastics, dyes and other chemicals are sometimes used in converting the x-ray effects to visible images that can be examined by the microscope. A complete review of such means is completely beyond the scope of this treatment. Suffice it to say that such recording or conversion materials thus far mentioned generally have resolutions of 1 μm or better, and so are compatible resolutionwise with the light microscope. Some materials are even essentially grainless, and no background structure can be detected even at the resolving limit of the light microscope.

Different specimen materials and the nature of the particular investigation determine the particular part of the x-ray spectrum that should be used. Selection of appropriate target materials and/or filters (such as absorption materials, grazing incidence reflectors and crystal reflectors) is pertinent to resolving this problem. Proper selection of recording materials also plays an important part in this aspect of x-ray microscopy.

In contact microradiography the microscope is utilized in several ways. When the x-ray image is recorded on film, which is then examined by the light microscope, the magnifications that may be used are over the full range of the instrument, including oil-immersion powers. Techniques include the full gamut of light microscope operation and adjustment, illumination, filters etc. Analysis of the microradiograph may even include such advanced techniques as high resolution microphotometry leading to micromass measurements of cells.

When the x-ray microfluoroscope is used (Figure 126) the microscope is able, via the converted x-radiation to fluorescent light, to view directly any motion or short interval changes in specimen conditions under radiation. Orientation of the specimen and visual selection of extended surfaces is more amenable using this method. It is in this form that perhaps the light microscope plays a more complete part in x-ray microradiography.

In contact microradiography as in x-ray microscopy in general, the primary considerations involving specimen preparation, resolution, contast and recording means, methods and materials are within the province of x-ray technique on a micro-level. The conventional light microscope is an adjunct to this process and whether it is used to examine x-ray exposed film, or whether it is incorporated into a microfluoroscope or other integrated x-ray instrument, the x-ray techniques themselves are primary to the final result. An extended treatment of x-ray technique in these cases is not possible in this brief discussion.

2. <u>Projection microradiography</u>: In this method the specimen is placed close to the x-ray source and a projected image obtained on a sensitive film material. The advantages include initial magnification as opposed to unit magnification in the contact method, and some improvement of the image to grain size ratio on the recording material. Even comparatively low magnifications of about 20X are very advantageous in this respect. Also, in addition, the ultimate magnification at which the microradiograph is examined by optical means is thereby effected, and may afford then more convenient optics use in the light microscope and less stringent demands upon its adjustment to obtain comparable or even better imaging of specimen details.

Some of the difficulties in the projection method include obtaining sufficient intensity at distances required for high magnification, and obtaining reasonable contrast in the image without encountering difficulties in sharpness of definition.

Figure 127 illustrates the essential parts of a projection x-ray microscope. Note that in contrast to the microfluoroscope of Figure 126 the specimen material is at the target-material point in this arrangement where the extremely small spot of x-radiation is formed by the concentrated (focused) electron beam impinging upon it. The x-radiation propagated through the specimen material over the distance from the specimen location to the film or fluorescent screen forms a primary enlarged specimen image that is directly related to that distance. Observation of the fluorescent screen can be by optical magnification using a microscope or the resultant magnified primary image on photographic film examined as is a contact microradiograph. Primary magnifications of up to 200X are obtainable with this type of instrumentation depending, of course, upon the specimen material and its characteristics.

3. <u>Electron microprobe</u>: In quality control, contamination control and sample and particle analysis, an electron microprobe is sometimes used. In addition to fundamental components, a specialized detector apparatus is used to analyze the x-radiation effected by the specimen material.

A practical device as such would include an electron gun and beam-forming system with a beam diameter of from 5 to 10 μm. Other essentials would include a mechanical stage and an optical microscope to determine which areas in the object are under examination, and a detector to assimilate and convert radiation for x-ray analysis.

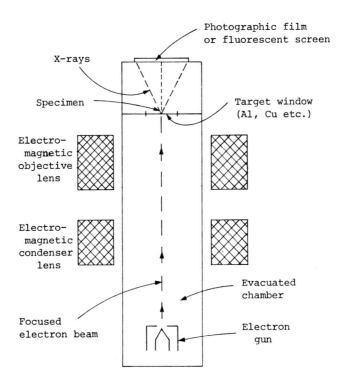

Figure 127. Essentials — schematic.
Projection x-ray microscope.

The sensitivity of the electron microprobe drops drastically for low-mass numbers and it cannot detect the three lowest at all. However, since in many routine areas of investigation, as those mentioned previously, analysis is restricted about 60 to 70% of the time to elements from magnesium upwards in the periodic table, this type of instrument finds wide use and application. The typical range of detection of elements with the electron microprobe is in the vicinity of from 10^{-12} to 10^{-14} grams.

Basically the instrument makes use of the specific x-radiation obtainable from different specimen materials. That is to say, the electron beam creates x-radiation from the impingement

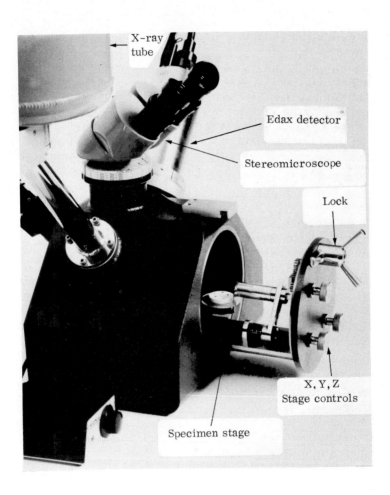

Figure 128. X-ray analysis microscope (MAX). (photo courtesy Edax North American Philips.)

upon specimen components rather than from a target material as in an x-ray tube, the specimen components themselves acting as targets. The resultant radiation upon being analyzed properly can reveal individual element presence and concentration to a very small degree.

A number of manufacturers of electron microscopes now provide accessory x-ray detection devices of the solid-state type for analysis of selected areas of the specimen so that electron microprobe work can be accomplished using the electron gun and beam forming system of the microscope.

Another type of instrument is represented by MAX, a device manufactured and marketed by EDAX, a North American Philips Company. It combines a light microscope for specimen observation and selection, an x-ray tube and generator, a vacuum chamber and a detector unit. A translating stage is included to position the specimen for features of interest for analysis. In addition to the detecting unit, an x-ray analyzer is necessary to accomplish the purpose of the instrument. Also, computers can be added to enhance qualitative analysis or provide quantitative results. Target materials to provide specific x-radiation include gold (Au), rhodium (Rh) and tungsten (W).

The microscope is binocular, has zoom optics (8 to 40X) (to 80X optional), and has an ocular reticle for indicating the analysis position. The specimen stage has a one-inch travel in X, Y and Z directions. Sample holders include low-background mylar film to support small particles, platforms for discs or pelletized samples and needle forceps.

This instrument can be used in identification, quantitative analysis, spectral analysis and comparison studies. The microscope, aside from its purely positioning-mechanical duties, often plays a vital part in the overall use of the instrument. Evaluation of color, fracture patterns, texture and other morphological features of specimen materials are sometimes very important in addition to x-ray analysis for identification and comparison work.

E. <u>ION MICROPROBE</u>: The ion microprobe is a mass spectrometer which examines a very small specimen sample.

Figure 129 shows in diagramatic form the elements of a mass spectrometer. Positive ions from a discharge tube pass through slits A and B and through an electric field between plates C and D and a magnetic field at E at right angles to the direction of travel. The electric field deflects the ions downward and the magnetic field deflects them upward, in both cases ions of higher

ION MICROPROBE

velocity being deflected less. The field intensities are adjusted so that, no matter what the ion velocities are, they will converge to a common focus on the photographic plate. Ions having different masses or charges will, of course, be deflected differently, and only those having the same charge to mass ratio (e/m) will arrive at a given point on the photographic plate. Where the ions have different masses they will appear at different points and consequently the atomic weights of them can be accurately determined.

In the ion microprobe, the ion source in Figure 129 is in actuality the specimen material under investigation. In other words, the specimen acts as a source of ions secondary in nature from an initial bombardment of it by an intense ion beam fabricated and focused especially for that purpose, i.e., to create secondary ions typifying the specimen material which are then analyzed by the mass spectrometer portion of the instrument.

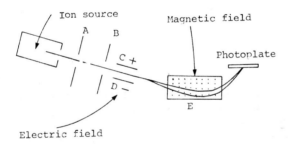

Figure 129. Elements of a mass spectrometer.

Figure 130 illustrates the essential elements of an ion microprobe. The primary ion source is usually obtained from a gas such as oxygen or argon, and the generated ions are focused and selected electrostatically and electromagnetically into a concentrated beam for bombarding the target (specimen) material. The specimen is observed and positioned with the use of an optical microscope and mechanical stage. The secondary ions sputtered off the specimen material are accelerated and focused into a mass spectrometer for analysis. The ion beam striking the specimen is focused onto an area approximately 1 μm in diameter. It has two major advantages over the electron micro-

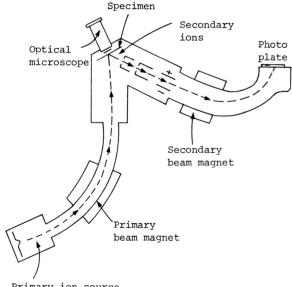

Figure 130. Essential elements of an ion microprobe.

probe. One, it is capable of encompassing the whole mass range of the periodic table, not being limited to the low-mass elements. Also, its sensitivity for most elements is in the range of 10^{-15} to 10^{-19} grams. Thus, this sensitivity of from 1 part per million (ppm) to 1 part per billion (ppb) on volumes of specimen material as low as a few hundredths of a cubic micrometer provides a very useful tool in trace analysis on a microscopic basis.

It is quite beyond the scope of this book to enter into detailed descriptions of the very complicated equipment involved. However, it is felt proper to point out that this sophisticated and sensitive scientific instrument is dependent upon the optical microscope to make it a practical instrument that can meet the demands of production and/or accelerated trace-analysis techniques. The ability to observe the specimen during analysis, and position specimen materials singly or in multiple, without breaking vacuum is a very great advantage. Aside from this "auxiliary" or assistant role, the optical microscope remains a full partner in the characterization of specimen material. Even though elemental analysis is carried out by this instrument, other

ION MICROPROBE

characteristics of the specimen material which may be just as important to the overall characterization can be, and are, obtained by the optical light microscope via techniques mentioned previously. A complete treatment of the various types of microprobes (electron, ion etc.) are included in other volumes of this series.

F. <u>LASER MICROSCOPE</u>: A coherent beam of high energy which can be trained on a selected location allowing any material, no matter how hard or tough, to be melted or vaporized is of great

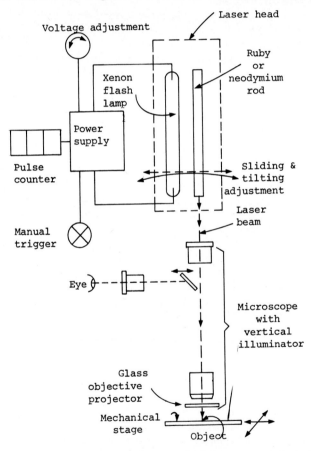

Figure 131. Essentials of a laser microscope. (adapted drawing courtesy Carl Zeiss Inc.)

interest engineering-wise and of wide ranging application to industrial purposes.

Combining the light microscope with a laser serves two purposes. First, the high magnification observation, location and positioning of materials in possible. Second, the optics used in reverse (from the ocular through the objective) provide a very fine focusing action on a laser beam.

The combination can provide for materials processing in the microrange (melting, welding, drilling, milling etc.) with great precision and reproducibility.

Carl Zeiss, Inc. has pioneered investigations into this area of laser-light microscope application and developed prototype equipment that is effective in materials processing.

The laser head contains a ruby or neodymium glass rod 150 mm long which is excited by a 2000 watt-second (joule) straight flash tube to pulses of about 1 millisecond duration at a laser energy up to 6 watt-seconds.

Cooling is by compressed air or in heavy duty-cycle operation water cooling is used. Ths laser head is mounted on a Zeiss Universal Microscope with mechanism allowing sliding and tilting adjustments in two dimensions. A slideable reflection system in the tube head of the microscope prevents simultaneous viewing during laser flashing. With short focal length objectives used at high magnifications a coverslip protector is placed between the specimen and objectives to prevent their damage by vaporized or ejected particles during lasing operations.

Intensity of the laser energy is adjustable by varying the amount of excitation by a xenon flash lamp, controllable either manually or automatically.

Industrial applications that appear to have immediate practical advantages are microwelding in microminiturization of electronic components, and drilling of diamond dies for wire production.

Aside from industrial applications, the laser microscope has considerable promise in the biological research field, especially in cell and tissue research. The effects of irradiation, localized heat and effects of strong electrical fields on cell research have only been touched upon.

In the medical area, the use of such instrumentation for therapeutic applications in the treatment of tumors, detached retinas, skin diseases and dental caries has been explored. The implications of this powerful tool for medicine alone are enormous.

LASER MICROSCOPE

Focusing a laser beam through a microscope and applying it in such concentration and intensity as to remove material for various fabrication purposes mentioned before has other implications as well. If the material sputtered off or otherwise vaporized can be collected and analyzed, a tool for ultramicrosampling is thus provided. Investigations into this area of application are being carried out currently. This type of "microprobe" will no doubt join other methods in providing a veritable battery of methods and techniques to more effectively characterize microscopic materials and specimens.

G. HOLOGRAPHIC MICROSCOPY: This method, originally conceived of to improve electron microscope resolution, has not fulfilled that promise but is making progress, if somewhat slow, in conjunction with the light microscope.

Basically, holographic microscopy consists of recording wavefronts from an object and reference field, on a photographic medium, and then reconstructing the object image from the recorded wavefront images and examining it (refer to Figure 132). Coherent light is used both in the recording and reconstruction steps of the method. The most convenient source of such light is from a laser beam. Both phase and amplitude information are recorded from the reference field and the object field on the recording medium. When reconstructed, by reinstituting a reference coherent light field that illuminates the recorded information, a real image will appear that has special properties, it appears to be three-dimensional, suspended in space.

The holographic microscope is an attempt to examine and obtain information from a microscope image of three dimensions. Only two different ways in which the light microscope has been used in this respect will be included here. In one method a holographic recording of a greatly magnified object by a microscope is viewed subsequently at little or no magnification. In the other, the microscopic object is holographically recorded at little or no magnification, and then the hologram is viewed and magnified by a light microscope.

The first method is illustrated in Figure 133. A laser furnishes both reference and object field illumination via a beam splitter. The reference beam is routed around the microscope via a mirror and another beam splitter at the recording surface. The magnified image at that surface is necessarily of a small field diameter. For technical reasons, a high magnification and large field diameter cannot be obtained at the same time with this

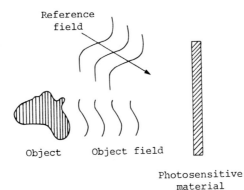

HOLOGRAPHIC RECORDING

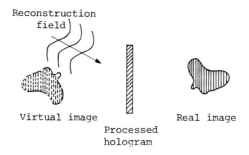

HOLOGRAPHIC RECONSTRUCTION

Figure 132. Two steps in holography. (illustration courtesy Microscope Publications Ltd.)

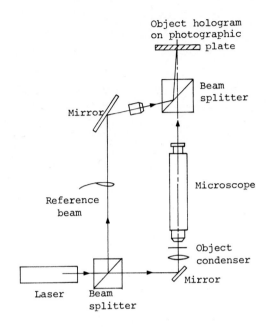

Figure 133. Holographic recording with a microscope. (illustration adapted courtesy Microscope Publications Ltd.)

method. The resulting hologram, upon reconstruction, is viewed at little or no magnification.

In the second method illustrated by Figure 134, the recording of the holographic image of a microscopic object is via "uncorrected" optics which are designed to provide a large diameter field, with an unavoidable sacrifice of image quality. Upon reconstruction, the recorded hologram produces a real image which is then further magnified and viewed by conventional microscope optics.

The advantages, disadvantages and other technical considerations of these two schemes and of many others that have been proposed and tried cannot be treated in this brief overview. It is, however, appropriate to indicate that there are many difficulties yet to be overcome in holographic microscopy, and although some

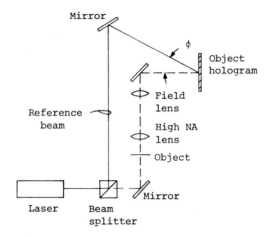

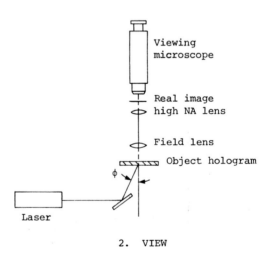

Figure 134. Holographic microscopy using "uncorrected" optics.

success has been achieved in very specialized situations and applications, a general application of the method remains for the future. Volume 27 of this series is entirely devoted to this fascinating subject.

H. SPECIALIZED RESEARCH INSTRUMENTATION

1. Introduction: It would not be possible in the limited space available to even briefly cover the innumerable research applications, integrations and adjunctal uses to which the optical microscope has been or is being put. Therefore, in the following paragraphs only a short selection of the ingenious or potentially more important research instrumentation is presented.

2. Optical research

a. Motorized photomultiplier scanner: Data Optics markets this device which is used to convert a stationary light distribution into a time varying electrical signal. The scanner includes a pinhole sampling aperture, a microscope focusing tube with a 10X objective and 5X ocular, and a photomultiplier detector. In operation, the pinhole is scanned across a stationary light distribution, producing a time-varying signal at the photomultiplier output which is proportional to the light distribution being scanned. Photomultiplier output is usually applied to a chart recorder for permanent records. The microscope is used to view the light distribution which is to be scanned, and for alignment of the pinhole aperture to be coincident with the object plane.

b. Optical bench microscope: This is a specially mounted general purpose instrument for optical research. Optics include objectives of 3.5X, 10X and 40X magnifications and a Leitz 6.3X Periplan ocular.

Micrometer adjustments of ±12 mm are provided for microscope motion along X, Y and Z axes. Two centers of rotation are provided, one around a vertical axis passing through the microscope body and another around a vertical axis in front of the microscope objective lens. The latter center of revolution (rotation) facilitates off-axis viewing of illumination in a horizontal plane. Description is courtesy Data Optics, Inc.

3. The cytopherometer: This Zeiss instrument is a special microscope designed for electrophoretic studies of cells and particles. If cells or some other kinds of particles (mineral grains) are suspended in a physiological fluid and then exposed to an electric field, migration toward the positive pole occurs. The velocity of the migration is dependent upon the sign and quantity

of the excess surface charge of the cell, and the velocity has been found to be a characteristic of certain cells, and useful in determination of pathological cell changes.

The instrument has two columns mounted on a base plate. One holds the electrophoresis system with condenser, and the other supports the microscope. The optical axis of the microscope is horizontal. The electrophoresis system consists of a measuring chamber inside a thermostatically controlled temperature chamber. Both chambers are mounted together on a rotating and centering stage. A lens is used as the observation window for the temperature chamber. It acts as the first element of a special microscope objective which has no physical contact with the measuring chamber. In this way, a relatively high-power objective is provided with a long-working distance. Magnifications as high as 800X are employed.

The illuminator is equipped with a brightfield phase contrast condenser with a numerical aperture of 0.6 and a low-voltage illuminator (6v, 15w) which projects from the rear of the column to allow dissipation of the heat.

Two electrodes are attached to the measuring chamber to maintain a constant electrical field inside it. Power for the purpose is supplied by an electronically controlled voltage regulator to the DC supply.

Photomicrographs can be taken via an available attachment camera that can be mounted between the microscope body and binocular tube.

Figures 135 and 136 illustrate the essential parts and their relative locations.

4. <u>Flow systems</u>: Instrumentation for cell analysis and sorting by high-speed, flow-through methods has been in use for some years. In a continuing state of development, there are a number of versions of such flow-system instruments in operation at present. Only one such instrument type will be briefly covered in these paragraphs to indicate principles and how the light microscope is associated with it.

The system is used to characterize cells by Coulter volume, light scatter and fluorescence. Coulter volume of a cell is directly related to a small change in electrical resistance which results from the passage of the cell (which is a poor electrical conductor) through a small orifice immersed in a conducting fluid such as physiological saline. Light scatter from individual cells results from the application of an incident argon-ion laser beam on

SPECIALIZED RESEARCH INSTRUMENTATION

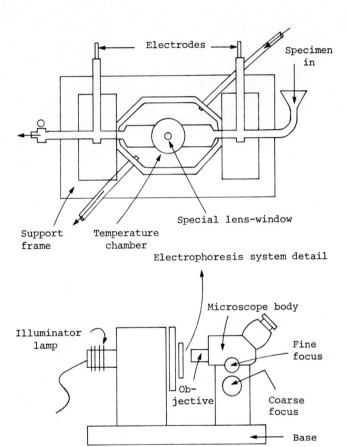

Figure 135. Cytopherometer details. (adapted drawing courtesy Carl Zeiss Inc.)

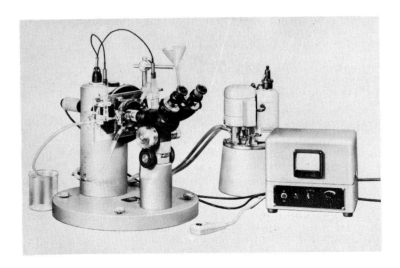

Figure 136. Cytopherometer. (photo courtesy Carl Zeiss, Inc.)

the cell as it passes through the flow chamber. When the cells are stained with certain fluorochromes, their degree of fluorescence due to illumination by the laser beam is also detected by a two-color fluorescence sensor.

The Coulter-volume electrical signal is amplified and sent to a multiparameter processing unit. The light-scatter amplitude is detected by a photodiode and further amplified before being connected to the processing unit. The fluorescent light is separated into red and green components by a dichroic filter and further channelized and converted into electrical signals which are also inputs to the processor. A very simplified functional drawing of the system is included in Figure 137.

The cells to be characterized are in a saline solution and made to pass through the flow chamber where they are illuminated one at a time by the blue light of the argon laser beam. As each cell passes through the beam it produces electrical (Coulter volume) or optical (light-scatter, fluorescence) pulses equal to the duration of cell transit time through the laser beam. Ampli-

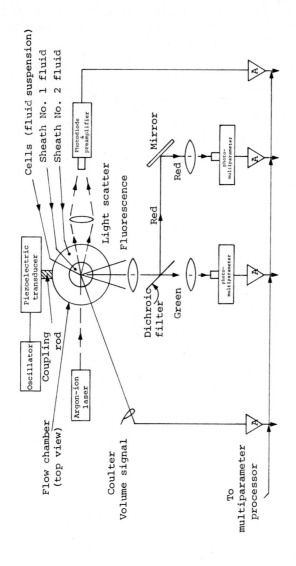

Figure 137. Multiparameter cell analysis system. (adapted drawing from illustration courtesy Los Alamos Scientific Laboratory.)

fication and introduction of these signals into an electronic processor then allows analysis on a statistical basis.

The mechanism whereby individual cells are isolated and made available for sorting on the basis of parameters indicated previously, is illustrated in Figure 138. The heart of the system is the flow chamber.

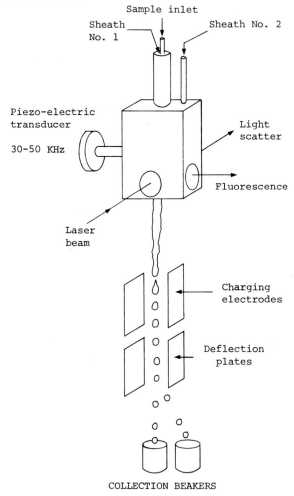

Figure 138. Multiparameter cell sorter operation. (adapted drawing courtesy Los Alamos Scientific Laboratory.)

The relative rates of sample and sheath no. 1 fluid flows are adjusted such that laminar flow results, permitting a cell stream of very fine diameter in much larger tubes (within the flow chamber), virtually eliminating clogging problems. Sheath no. 1 flow is laminar around the sample inlet tube and thereby serves to center coaxially the cell stream as both flow through a volume-sensing orifice where Coulter volume is measured. Upon exiting from the volume-sensing orifice the cell stream-sheath no. 1 flow enters another fluid filled region (sheath no. 2) and here is exposed to the laser beam for scatter-signal generation and/or fluorescence production from appropriately fluorochromed cells. Sheath no. 2 fluid flows coaxially around the cell stream-sheath no. 1 flow, facilitating droplet formation. Excitation of the cell stream-sheath no. 1-sheath no. 2 flow, by a piezo-electric transducer mechanically coupled to the flow chamber generates uniform droplets. Cells are effectively isolated in this manner inside single liquid droplets. Individual droplets are generated at a rate of about 45,000 per second.

By establishing preset signal criteria on one of the parameters (Coulter volume, light scatter or fluorescence) the droplets meeting the criteria are charged electrically and deflected accordingly into a separate collection vessel. This sorting action can take place at a rate of about 200 to 300 cells per second.

The light microscope plays three important parts in flow-system analysis. First, a stereomicroscope is used within the instrumentation assembly of the system for adjustment purposes. Observation of droplet formation during adjustment of the system and while it is processing samples is an important factor in reliability and repeatability of results. Also, a conventional light microscope is used in the checking of the sorting operation on a spot-check basis using either in-process cells or test glass-microspheres.

Sample preparation involves disaggregation of cellular tissue into single cell entities, and fixing and staining procedures of a complex and delicate nature. The light microscope in many different versions (including phase contrast, fluorescence etc.) is an essential instrument in assuring proper specimen preparation.

After the sample has been processed and data obtained from electronic processing and computing devices, confirmation and correlation is performed on sorted cells with the light microscope.

As mentioned previously, there are a number of versions of flow-system devices, all mainly devoted to the analysis of mammalian cells. However, it has been suggested that analysis and sorting of other types of cells and/or particles might also be feasible using similar flow-system principles.

The separation, identification and/or characterization of individual cells at the tremendous rates possible with this type of instrumentation is advancing cancer and other cell research impressively. Extension to other areas of research in the future seems certain.

5. <u>Visual stimulation apparatus (VSA)</u>: The VSA is an example of very special adaption of light microscope optics to a physiological research device. It is a projection instrument with two independent beam paths which project slides on the concave side of a hollow sphere via a single small exit pupil which coincides with the surface of a rotatable and tiltable mirror. According to the position of the mirror a luminous field (stimulus field) is produced at one point or other of the projection surface within which patterns (slide images) are moved. The VSA is largely parallax free because the small exit pupil only requires a small rotatable mirror which is positioned in the center of the projection sphere near the subject's eye.

The instrument is used to produce visual stimuli for the purpose of analysis by electrophysiological methods (brain probe) of the visual system of vertebrates.

The nucleus of the imaging optics are two Zeiss Standard microscopes each with a 2.5X/0.08 NA planachromat objective. The stage of each microscope carries an individual test pattern. By a complex system of lens elements and beam splitters a mechanically modified C5 (Zeiss) ocular produces a final image at a total magnification of 12.5X and at a field angle of 23° in the projection sphere (300 mm radius with interior matte white surface).

The exit pupil, pattern position and orientation, and location of the projected image on the projection sphere are all controllable by electromechanical drives. A very simplified schematic of the essential features of the instrument is shown in Figure 139. The many refinements of operation control and detailed application are beyond the scope of this brief treatment.

I. <u>REFERENCES AND COMMENTARY</u>

1. Knox, C. and R. E. Brooks, "Holographic motion-picture microscopy," <u>J. S.M.P.T.E.</u> 79, 594-8 (1970).

2. Simpson, J. A., "Use of a microscope as a non-contacting microdisplacement measurement device," <u>Rev. Sci. Instr.</u>

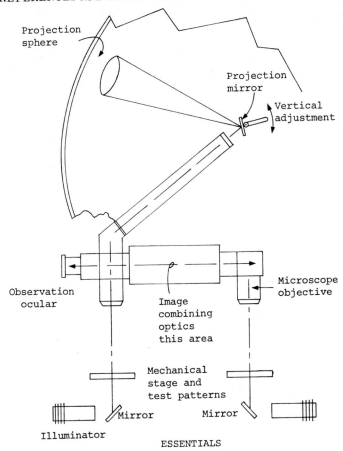

Figure 139. Visual stimulation apparatus. (adapted drawing courtesy Carl Zeiss, Inc.)

$\underline{42}$ (9), 1378-80 (1971).

3. Black, J. F., C. V. Summers and B. Sherman, "Scanned-laser microscope for photoluminescence studies," Appl. Opt. $\underline{11}$, 1553 (1972).

4. Cox, M. E., R. G. Buckles and D. Whitlow, "Cine-holo microscopy of small animal micro-circulation," J. Opt. Soc.

Am. 59, 1545 (1969).

5. Jeong, T. H. and H. Snyder, "Holographic microscope system using a triangular interferometer," Appl. Optics 12, 146 (1973).

6. Van Ligten, R. F., "Holographic microscopy," J. Opt. Soc. Am. 60, 709, (1970).

7. Schindl, K. P., "A new bore-hole inspection microscope," The Microscope 20, 183 (1972).

8. Adams, M.D., "Ultramicrosampling with a laser microscope," The Microscope 19, 157-69 (1971).

9. Davis, A.M., "Contact microradiography in the study of alloys," Metallography 3, 165-82 (1970).

10. Mela, M. J. and M. S. Sulonen, "A new method of laser-microprobe analysis," J. Phys. E. 3, 901-3 (1970).

11. White, G. W., "Improving the accuracy of vertical measurements under the microscope," The Microscope 18, 51-59 (1970).

12. Anderson, W. L., R. E. Beissner and R. L. Bond, "High resolution large volume microscopy with low spatial-frequency hologram," J. Opt. Soc. Am. 60, 715 (1970).

13. Leith, E. N. and J. Upatnieks, "Microscopy by wavefront reconstruction," J. Opt. Soc. Am. 55, 569 (1965).

14. Sakayanagi, Joshimi, "Sharpness of an edge in microscopic measurement," J. Opt. Soc. Am. 60, 1530-1 (1970).

15. Mullaney, P. F., J. A. Steinkamp, H. A. Crissman, L. S. Cram and D. M. Holm, "Laser flow microphotometers for rapid analysis and sorting of individual mammalian cells," Biophysics and Instrumentation Group, Los Alamos Scientific Laboratory, Los Alamos, New Mexico.

16. Salzman, G. C., J. M. Crowell, C. A. Goad, K. M. Hansen, R. D. Hiebert, P.M. LaBauve, J. C. Martin, M. L.

Ingram and P. F. Mullaney, "A flow-system multiangle light-scattering instrument for cell characterization," Clinical Chemistry 21, 1297 (1975).

17. Clark, (Ed.), The Encyclopedia of Microscopy, (Section on X-ray Microscopy) pp. 561-593, 1961. A very good general treatment of various aspects of the use of x-rays at the microlevel. Contains many reference listings in specialized areas.

18. Cox, Mary E., "Holographic microscope — a reassessment," The Microscope 22, 4th Quarter (1974).

19. Asunmaa, Saara K., "Improved resolution in x-ray absorption micrographs," The Microscope 14, (2) (September-October 1963); Part I, Part II, 14, (3), (November-December 1963).

20. Ely, R. V., "X-ray microscopy with electronic magnification," The Microscope 14, (11), (July-August 1965).

21. Cox, M. E., "Holographic microscopy — a review," The Microscope 19, 2nd Quarter (1971).

22. Nelson, J. B., "Characterization of materials by x-ray microscopy," Part I, The Microscope 19, 4th Quarter (October 1971).

23. Keller, H. E., "Quantitative determination of surface topography by light microscopy," The Microscope 21, 1st Quarter (January 1973).

24. Neupert, Helmut, The Cytopherometer, Zeiss Information No. 60, Production News, 1966.

25. Precision Measuring Microscope, with Eyepiece Screw Micrometer, Zeiss Information No. 59, Production News, 1965.

26. Kleinsasser, Oskar, Endolaryngeal Microscopy and Photography, Zeiss Information No. 60, 1966.

27. Vohse, Gerhard, New Toolmakers Measuring Microscope, Zeiss Information No. 63, 1966.

28. Jakubowski, H. and H. Reidel, The Motorized Head for Electric Vertical Movement of the Operation Microscope, Zeiss Information No. 63, Production News, 1966.

29. Panzer, Siegfried, Microscope and Laser, Zeiss Information No. 64, 1967.

30. Marsch, Angelika, The Ultraphot II as Reflected-Light Microscope in Metallography, Zeiss Information No. 64, 1967.

31. Kinder, Walter, The Interference Flatness Tester and Instruments for Surface Testing, Zeiss Information No. 58, 1965.

32. Littmann, Hans, The New Zeiss Slit Lamp, Zeiss Information No. 58, 1965.

33. Littmann, Gert and Reimar Wittekindt, Operation Microscope with New Camera Attachment and New Observation Tube for a Second Observer, Zeiss Information No. 58, 1965.

34. Vohse, Gerhard, Rapid Alignment Fixture for UMM Universal Measuring Microscope, Zeiss Information No. 65, 1967

35. Reidel, Helmut, Heinz G. Jakubowski and Gunther Summerer, Assistants Microscope and Operating Field Magnifier, Zeiss Information No. 79, 1973.

36. Reinig, Hans-Joachim, Visual Stimulation Apparatus, Zeiss Information No. 82, 1973/74.

37. Measuring Microscopes for Laboratory and Shop, Gaertner Scientific Corporation, Bulletin 161-75.

SUBJECT INDEX

Abbe condenser, 26
Abbe test-plate, 69, 70
absorption, 187
achromat, 4
aids to counting, 233
anoptral phase contrast, 107
apochromat, 5
aqueous media, 145
areal and linear analysis, 251
artifacts, 128
    cutting and grinding, 131
    Mach effect, 64
    optical, 62, 63
    phase contrast, 63
    projection, 63
    surface preparation, 163
astigmatism, 64
automated image analysis
    basic system, 264, 265
    data results, 268
    display and presentation, 270
    field measurements, 268
    multiple field measurement, 270
    processing, 265
    programming, 267
    scanning, 264
    selected measurements, 268
    selective particle measurements, 269
    stage and search patterns, 267
    switch controls, 266
    system control, 266
automated photomultiplier scanner, 317
automatic point counter, 261
biaxial interference figure, 196
bi-chromatic illumination, 286
binocular microscope, 59

box mounts, 150
bubbles in mountants, 156
calibration liquid, 148
casting, 136
cathetometer, 278
choosing optics, 3-26
chromatic aberration, 72
circular polarization, 188
coma, 64
comparators, 293
comparison of oculars, 20
compensating ocular, 21
condenser(s)
    cardioid, 90
    centering, 37
    comparison, 28
    corrected, 27
    darkfield, 88
    incident light, 92
    light cones, 39
    paraboloid, 90
    ring test, 40
    types, 27
    vertical adjustment, 38
contact microradiography, 301, 302
contrast
    color, 84
    improve, 81-126
    modulation, 109
    sensitivity, 83
correction collars, 46
corroding, 130
counting analysis
    automatic accessories, 260
    by convention, 248
    chart, 241
    compositional, 250
    field plans, 245, 246
    fields, 242
    magnitude, 242
    procedures, 241
    quality, 250

counting analysis (continued)
   search patterns, 244
   slide preparation, 247
   tabulation form, 241
counting chamber
   blood, 236
   Dunn, 235
   Howard, 235
   Helber, 235
   Petroff-Hauser, 235
   Sedgewick-Rafter, 236
coverslip
   effects, 47
   spherical aberrations, 43
   thickness, 41
   thickness effect, 42
   thickness limits, 44
cutting
   animal and plant tissues, 128
   bone, 129
   fibers, 129
   other materials, 129
   wood, 129
darkfield
   condensers, 88
   illumination, 87
   practical matters, 89
   stops, 87
depth measurement, 289
diatom test slides, 73
differential count recorder, 262
differential interference contrast, 114, 116, 170
   Nomarski, 118, 119, 121
diffraction, 186
   effects, 62
dispersion, 186, 218
   curves, 216
dispersion staining, 116
   characteristic indicators, 221

dispersion staining (continued)
   characteristics determined, 221
   colors, 222
   darkfield methods, 217, 219
   objective, 220
   objective light path, 221
   special considerations, 222
double counting, 247
dry mounts, 150
electron microprobe, 305
equipment adjustment, 30
etching, 134
examination of periodic structures, 56
excitation, 190
extensometers, 293
eye, spectral sensitivity, 85
field of view, 68
filters
   absorption and interference, 96
   applications, 97
   colored, 93
   contrast, 96
   exciter and barrier, 205
   guide for contrast, 97
   infrared, 100
   interference, 93
   Kodak Wratten, 98, 99
   light, 32
   liquid, 100
   neutral density, 34
   polarized light, 101
   resolution improving, 35
flare, 82
flatness of field, 68
flow systems, 318
fluid mounting
   electrified, 163, 164
   media, 148
   temporary, 159

fluorescence microscope, 202, 203, 206
fluorescence microscopy, 172, 202
  characteristic indicators, 204
  characteristics determined, 204
glare, 82
glass 173
grinding and polishing, 133
grinding rocks and minerals, 130
Hartshorne rotation apparatus, 214
heavy liquids, 239
Helber counting chamber, 235
Hoffman modulation contrast, 108
hole-inspection microscope, 291, 292
holographic microscopy, 313, 316
  recording, 315
Howard counting chamber, 235
illumination
  coherent, 83
  darkfield, 87
  incident, 91
  intensity, 32
  Köhler, 30, 31, 33
  oblique, 87
  operation microscope, 297
  phase contrast, 106
  relative lighting, 60
  Rheinberg, 88
  symmetry of, 68
immersion cap, 166
immersion fluids, 145
immersion oil, 51, 148
  characteristics, 53
index of refraction liquids, 148
industrial operations, 273
infrared microscopy, 171
intensity contrast, 82

interference, 186
interference contrast, 108
interference microscope
  A-O Baker, 111, 113, 115
  Zeiss, 112
interference microscopy, 170, 197
  characteristic indicators, 198
  characteristics determined, 198
  double-focus system, 114
  film thickness by, 201
  measurements by, 197
  morphology by, 197
  multiple-beam, 202
  shearing system, 114
  structural analysis by, 198
  systems compared, 115
inverted microscope, 177, 178
ion microprobe, 308, 310
iris diaphram, 86
Jamin-Lebedeff system, 110
Joy-stick control, 266
killing and fixing, 127
laser microscope, 311
light characterization phenomena, 184
light filters, 32
light, incident and reflected, 224
light-pen control, 266
liquid filters, 32
mass spectrometer, 309
measuring microscope, 274
  types of measurements, 275
medical practice, 294
metallization, 136
micropaleontological slides, 151, 153
microphotometry, 224
  characteristics determined, 227

Microphotometry (continued)
  essentials, 226
  instrumentation, 225
microscope
  cleanliness, 60
  parameter measurements, 65
  performance testing, 69
  range, 3
  testing, 65
  total magnification, 25
microscopy
  as adjunct, 273-328
microspectrophotometry, 229
mountants, 137
  index of visibility, 141
  list of, 145, 146, 147
  low index, 145
  properties, 138, 142
  penetration by, 156
  refractive index, 138, 140
  temporary, 148
mounting methods, 149
moving microscope
  fixed-stage, 277
multiparameter cell analysis
  system, 321
  cell sorter, 322
numerical aperture
  relationships, 10
  defined, 7
objective, 3
  achromat, 4
  apochromat, 5
  back focal plane, 56, 58
  barell notations, 14
  chromatic aberration, 72
  comparisons, 7
  correction collars, 46
  depth of field, 12, 67
  equivalent focal length, 17
  flatness of field, 68
  fluorite, 4, 5
  high-dry, 17

objective (continued)
  immersion, 5
  infinity corrected, 6
  intermediate image distance, 15
  lens arrangements, 22
  numerical aperture measurements, 65
  object distance, 15
  optical-mechanical relationships, 15
  planachromat, 4
  polarization of, 68
  resolution vs NA, 8
  semi-apochromat, 5
  working distance, 15, 67
ocular(s), 19
  comparison, 20, 21
  compensating, 21
  Huygenian, 20
  Kellner, 21
  lens arrangements, types, 22
  orthoscopic, 276
  Ramsden, 21
  Ramsden disc, 23
  Ramsden disc determination, 67
operation microscope, 294
  floor mounted, 296
  with camera attachment, 295
ophthalmology, 297
optical artifacts, 61
optical-bench microscope, 317
optical research, 317
particle geometry, 257
  average dimension, 258
  globe and circle, 258
  Martin's diameter, 258
  measurement terminology, 269
  short dimension, 258
  size and shape, 258

particle geometry (continued)
   two-tangent distance, 258
particle size analysis reticles, 259
permanent mounts, 166
phase contrast
   basic system, 102
   bright-medium mode, 169
   centering annuli, 103
   dark-medium mode, 169
   dark-low-low mode, 169
   microscope, 104, 105
   objective, 103
   positive and negative, 106
   practical considerations, 103
   special considerations, 168
photometry vs stereometry, 228
photo-slit-lamp
   with beam splitter, 298
   Zeiss, 301
point counting, 253
   point spacing, 255
   error estimation, 256
polarization, 187
polarization microscope, 174
   essentials, 192
   research type, 194
polarization microscopy, 171, 190
   characteristic indicators, 193
   characteristics determined, 193
portable microscope, 174
   precision, 176
   McArthur O.U., 176
projection microradiography, 305
quartz wedge, 197
Ramsden ocular, 21
references
   adjunctal microscopy, 324

references (continued)
   counting and image analysis, 271
   improving contrast, 123
   improving resolution, 76
   sample characterization, 230
   specimen preparation and observation, 179
reflection, 183
reflectivity, 227
refraction, 183
replication
   film impression, 135
   gummed tape, 135
resinous mounts, 155
resolution
   definition, 1
   methods to improve, 1-80
   numerical aperture, 8
   summary to improve, 76
resolving power formula, 1
reticles, 233
   measuring, 281
   grain size, 252
   magnification, 284
ringing turntable, 158
Rousellet compressor, 165
sample characterization, 183-231
   crystal habit, 212
   crystal rolling, 213
   dispersion staining, 216
   form and orientation, 211
   fluorescence, 202
   interference, 197
   microphotometry, 224
   microhardness, 208
   microspectrophotometry, 229
   polarization, 191
   refractive index, 185
   ultraviolet, 223

sampling, 233
Schlieren microscopy, 121
semi-apochromat, 5
sieves, 238
slit-lamp, 298, 299
special observing conditions, 174
specialized research instrumentation, 317
specimen examination
  darkfield, 168
  differential interference contrast, 170
  fluorescence microscopy, 172
  infrared microscopy, 171
  interference microscopy, 170
  phase contrast, 168
  polarization microscopy, 171
  ultraviolet microscopy, 172
specimen preparation
  casting, 136
  corroding, 130
  cutting, 128, 129, 130
  etching, 134
  grinding, 130
  grinding and polishing, 133, 154, 155
  killing and fixing, 127
  metallization, 136
  mounting, 149
  replication, 135
  sectioning, 128
  semi-embedding, 136
  staining, 131
spherical aberration, 42, 64
substage condensers, 26, 37
substage diaphram control, 41, 86
surface examination, 133
surface finish, 289
temporary culture slides, 162
temporary mounts, 159

test objects, 68
test slides
  Abbe, 69, 70
  bacterial flora smear, 73
  diatoms, 73, 75
  human blood, 73
  proboscis of a blow-fly, 73
  star-test, 69, 70
  wing of dragonfly, 69, 70
toolmaker's microscope
  accessories, 286
  column tilt, 282
  digital readout, 287
  filar micrometer ocular, 285
  lighting, 285
  optical system, 279
  reticle types, 282
  surface index blocks, 287
  with depth gauge, 290
transmissibility, 228
transparency medium, 148
tubelength
  corrector, 50
  general comments, 50
ultramicroscopy, 92
ultrasonic devices, 239
ultrasonic sifter, 240
ultraviolet, 35
ultraviolet microscopy, 172, 233
universal stage, 215
use of back focal plane, 54
use of filters, 93
viewing-field diameter, 18
visibility, 81
visual bandpass filter characteristics, 94
visual depth of field, 11
visual Mach effect, 64
visual stimulation apparatus (VSA), 324, 325
working conditions, 60
x-ray analysis microscope (MAX), 307

x-ray microfluoroscope, 303
x-ray microscopy, 300
   projection, 303
   spectra, 303
Zeiss interference microscope, 111
Zeiss slit-lamp, 301

AUTHOR INDEX

Adams, M. D., 326
Allen, R. M., 181
Anderson, W. L., 326
Arkin, H., 271
Asunmaa, S. K., 327
Baker, J. R., 78
Barer, R., 80
Barnett, M. I., 80
Barnett, W. F., 179
Barnett, W. J., 181
Bayard, M., 271
Beissner, R. E., 326
Bell, A. S., 78
Belling, J., 77
Bhatnagar, G. S., 79
Black, J. F., 325
Bond, R. L., 326
Brooks, R. E., 324
Brown, J. F. C., 272
Brown, K. M., 230
Buckles, R. G., 325
Butler, J., 271
Butler, M., 271
Cameron, E. N., 181
Cargille, J. J., 80
Clark, G. L., 125, 181, 327
Cole, M., 271, 272
Colton, R. R., 271
Condon, E. U., 229
Conrady, A. E., 78
Cox, M. E., 325, 327
Cram, L. S., 326
Crissman, H. A., 326
Crossman, G. C., 179
Crowell, J. M., 326
Dade, H. A., 179, 180
Davis, A. M., 326
Defleur-Schenus, M., 182
Dehoff, R. T., 230
Delly, J. G., 79, 181
de Ternant, P., 179
Dimersoy, S., 230
Dodd, J. G., 125

Ely, R. V., 327
Evens, E. D., 179, 180
Fisher, C., 271, 272
Fleming, W. D., 182
Ford, W. E., 230
Forlini, L., 230
Frison, E., 179, 180
Furman, N. H., 230
Gahn, J., 230
Geissinger, D., 79
Goad, C. A., 326
Goodman, R. A., 230
Gray, P., 181
Gross, L. J., 126
Grosskopf, R., 272
Hanna, G. D., 181, 182
Hansen, K. M., 326
Hartridge, H., 78
Hartshorne, N. H., 230
Hashimoto, H., 126
Hausmann, E., 229
Herbosch, A., 182
Heunert, H-H., 182
Hiebert, R. D., 326
Hoffman, R., 126
Hohn, D. M., 326
Humphries, D. W., 272
Ingram, M. L., 327
Jackson, A., 125
Jakubowski, H., 328
Jedwab, J., 182
Jenkins, F. A., 125
Jeong, T. H., 326
Jesse, A., 272
Jones, F. T., 77, 230
Keller, H. E., 327
Kinder, W., 328
Kirkpatrick, A. F., 77
Kleinsasser, O., 327
Knox, C., 324
Kumeo, A., 126
LaBauve, P. M., 326
Lambert, W. E., 80